D1032996

Lavery Library
EX LIBRIS
ST. JOHN FISHER COLLEGE

Topics in Number Theory

THE UNIVERSITY SERIES IN MATHEMATICS

Series Editor: Joseph J. Kohn
Princeton University

THE CLASSIFICATION OF FINITE SIMPLE GROUPS
Daniel Gorenstein
VOLUME 1: GROUPS OF NONCHARACTERISTIC 2 TYPE

ELLIPTIC DIFFERENTIAL EQUATIONS AND OBSTACLE PROBLEMS
Giovanni Maria Troianiello

FINITE SIMPLE GROUPS: An Introduction to Their Classification
Daniel Gorenstein

INTRODUCTION TO PSEUDODIFFERENTIAL AND FOURIER INTEGRAL OPERATORS
François Treves
VOLUME 1: PSEUDODIFFERENTIAL OPERATORS
VOLUME 2: FOURIER INTEGRAL OPERATORS

MATRIX THEORY: A Second Course
James M. Ortega

A SCRAPBOOK OF COMPLEX CURVE THEORY
C. Herbert Clemens

TOPICS IN NUMBER THEORY
J. S. Chahal

Topics in Number Theory

J. S. Chahal
Brigham Young University
Provo, Utah

Plenum Press • New York and London

Library of Congress Cataloging in Publication Data

Chahal, J. S.
Topics in number theory / J. S. Chahal.
p. cm.—(The University series in mathematics)
ISBN 0-306-42866-0
1. Numbers, Theory of. I. Title. II. Series: University series in mathematics (Plenum Press)
QA241.C52 1988
512′.7—dc19 88-4126
CIP

A Division of Plenum Publishing Corporation
233 Spring Street, New York, N.Y. 10013

Printed in the United States of America

To Professor John T. Tate

Preface

This book reproduces, with minor changes, the notes prepared for a course given at Brigham Young University during the academic year 1984–1985. It is intended to be an introduction to the theory of numbers. The audience consisted largely of undergraduate students with no more background than high school mathematics. The presentation was thus kept as elementary and self-contained as possible. However, because the discussion was, generally, carried far enough to introduce the audience to some areas of current research, the book should also be useful to graduate students. The only prerequisite to reading the book is an interest in and aptitude for mathematics. Though the topics may seem unrelated, the study of diophantine equations has been our main goal.

I am indebted to several mathematicians whose published as well as unpublished work has been freely used throughout this book. In particular, the Phillips Lectures at Haverford College given by Professor John T. Tate have been an important source of material for the book. Some parts of Chapter 5 on algebraic curves are, for example, based on these lectures. The chapter on the computation of Mordell–Weil groups is borrowed from his lectures without any changes. Siegel's proof of Dirichlet's theorem on the group of units of an algebraic number field is from a course given by Professor Takashi Ono at the Johns Hopkins University (with the exception that we have avoided using finiteness of the number of ideals of bounded norm). The proof of the Mordell–Weil theorem is from Weil's paper of 1930, "Sur un Théorème de Mordell." The elementary proof of the "Riemann hypothesis" is due to Yuri I. Manin. An important but not well explained argument in Manin's original paper has been clarified. The main and excellent source of our information on finite fields has been the lectures on "Equations over Finite Fields" by Professor Wolfgang M. Schmidt.

I would like to thank Professor Schmidt and Professor Tate for suggesting several improvements in the manuscript.

Finally, I would like to express my gratitude to Professor Stephen P. Humphries for his help in proof reading and to Lonette Stoddard and Jill Fielding for the excellent job of typing the manuscript.

Theorems marked with an asterisk have not been proved in this book. The interested reader can find those proofs in the references we have cited.

After the book had been completed, several texts appeared that provide excellent material for further reading:

D. Husemoller, *Elliptic Curves*, GTM 111, Springer Verlag, New York (1987).
N. Koblilz, *Elliptic Curves and Modular Forms*, GTM 97, Springer Verlag, New York (1985).
J. H. Silverman, *The Arithmetic of Elliptic Curves*, GTM 106, Springer Verlag, New York (1986).

J. S. Chahal

Provo, Utah

Contents

Notation

$\varnothing$	The empty set
$X \cup Y$	The union of two sets X and Y
$X \cap Y$	The intersection of X and Y
$X \subseteq Y$ or $Y \supseteq X$	X is a subset of Y
$X \subsetneqq Y$ or $Y \supsetneqq X$	X is a proper subset of Y
$x \in X$ or $X \ni x$	x is in X
$x \notin X$ or $X \not\ni x$	x is not in X
$\|X\|$	The number of elements in X
$\{x \mid P(x)\}$	The set of all x having the property $P(x)$
$X - Y$	$\{x \in X \mid x \notin Y\}$
$f: X \to Y$	f is a (map or) function from X into Y
$g \circ f$	Composition of functions
$X \ni x \to y \in Y$	A function taking x in X to y in Y
$\mathbb{N} = \{1, 2, 3, \ldots\}$	The natural numbers
$\mathbb{Z} = \{0, \pm 1, \pm 2, \ldots\}$	The integers
$\mathbb{Q} = \{m/n \mid m, n \in \mathbb{Z},\ n \neq 0\}$	The rational numbers
$\mathbb{F}_q$	A finite field of q elements
$\mathbb{R}$	The real numbers
$\mathbb{C}$	The complex numbers
$\mathbb{H}$	The Hamiltonians
$[x]$	The integer part of a real number x
$\exp(x)$	The exponential function e^x
$\|z\|$	Absolute value of a complex number z
$\operatorname{Re} z$	Real part of z
$\operatorname{Im} z$	Imaginary part of z

$A[x_1, \ldots, x_n]$	The polynomials in n variables $x_1, \ldots, x_n$ with coefficients in a ring $A = \mathbb{Z}, \mathbb{Q}, \mathbb{F}_q, \mathbb{R}, \mathbb{C}$, etc.
$M(n, A)$	The ring of $n \times n$ matrices over a ring A
$A^\times$	The group of units of a ring $A \ni 1$
P'	the transpose of a matrix P
$\|P\|$ or $\det(P)$	The determinant of P
$\mathrm{GL}(n, A)$ or $\mathrm{GL}_n(A)$	$\{P \in M(n, A) \mid \det(P) \in A^\times\}$
$\mathrm{SL}(n, A)$ or $\mathrm{SL}_n(A)$	$\{P \in \mathrm{GL}(n, A) \mid \det(P) = 1\}$
$\mathbb{E}(L)$	Elliptic functions with period lattice L
$\Rightarrow$	Implies
$\Leftarrow$	Is implied by
$\Leftrightarrow$	If and only if
$\sum_{j=1}^{n} a_j$	The sum $a_1 + \cdots + a_n$
$\prod_{j=1}^{n} a_j$	The product $a_1 \cdots a_n$

1

Basic Properties of the Integers

1.1. Divisibility

The most fundamental concept in the study of the integers is that of divisibility.

DEFINITION 1.1. If $a, b \in \mathbb{Z}$ and $a \neq 0$, we say that a *divides* b or that a is a *divisor* of b if $b = ac$ for some c in $\mathbb{Z}$. We write it as $a \mid b$.

If $a \mid b$ we also say that a is a *factor of* b or that b is a *multiple of* a. As an example, $2 \mid 6$, because $6 = 2 \cdot 3$. If a does not divide b, we write it as $a \nmid b$.

An integer is called *even* if it is a multiple of 2; otherwise it is *odd*. Any odd integer a can be written as $a = 2m + 1$ with m in $\mathbb{Z}$.

The following theorem is obvious from the definition of divisibility.

THEOREM 1.2.

1. *$a \mid a$ for any $a \neq 0$ in $\mathbb{Z}$.*
2. *If $a \mid b$, and $b \mid a$, then $b = \pm a$.*
3. *If $a \mid b$, then $a \mid bc$ for any c in $\mathbb{Z}$.*
4. *If $a \mid b$ and $b \mid c$, then $a \mid c$.*
5. *If $a \mid b$ and a, b are both positive, then $a \leq b$.*
6. *If $d \mid a$ and $d \mid b$, then $d \mid ax + by$ for any pair of integers x and y. In particular, $d \mid 0$ for any nonzero integer d.*

1.2. The Division Algorithm

We can divide an integer by a nonzero integer to get a quotient and remainder. More precisely, we have the following theorem.

THEOREM 1.3 (*The Division Algorithm*). *Given n and d in $\mathbb{Z}$ with $d \geq 1$, there are unique integers q and r $(0 \leq r < d)$ such that $n = qd + r$. In particular, $q \mid n$ if and only if $r = 0$.*

PROOF. Because the real line $\mathbb{R}$ is the disjoint union of the semiclosed intervals

$$I_j = [jd, (j+1)d) = \{x \in \mathbb{R} \mid jd \leq x < (j+1)d\},$$

where $j = 0, \pm 1, \pm 2, \ldots$, the integer n is in a unique interval I_q. Since each interval I_j is of length d, $0 \leq n - qd = r < d$. □

1.3. Primes

DEFINITION 1.4. An integer $p \geq 2$ is called a *prime number* or simply a *prime* if ± 1 and $\pm p$ are the only divisors of p.

The first few primes are

$$2, 3, 5, 7, 11, 13, 17, \ldots.$$

The primes occur "at random."

An integer $n > 1$ is called a *composite number* if it is not a prime. Thus a composite number n is a product of two strictly smaller integers, each of which, if not a prime, is further a product of strictly smaller integers. Because this process cannot continue forever, we arrive at an expression for n as a product of primes. Thus we have proved the following theorem.

THEOREM 1.5. *Every integer $n > 1$ is a product of primes.*

Let $n > 1$ be a composite number. We write $n = d_1 d_2$, $1 < d_i < n$ $(i = 1, 2)$. Without loss of generality, let $d_1 \leq d_2$. If p is a prime divisor of d_1, then $p^2 \leq d_1^2 \leq d_1 d_2 = n$, which shows that n has a prime divisor $p \leq \sqrt{n}$. This provides us with a process of determining all the primes between 1 and N. It is called the *sieve of Eratosthenes* and consists of deleting from the list

$$2, 3, 4, 5, 6, 7, 8, \ldots, N$$

all the nontrivial multiples first of 2, then of 3, then of 5, and so on. At any stage $j > 1$ (starting with $p_1 = 2$) the first integer $p_j > p_{j-1}$ to survive the deletion of the nontrivial multiples of p_{j-1} is the jth prime. By what has been said above, it suffices to delete the nontrivial multiples of all the primes $p \leq \sqrt{n}$ only.

EXERCISE 1.6. Use the Sieve of Eratosthenes to list all the primes less than $N = 400$.

Every positive integer can be factored in one and only one way as a product of primes. This is called "*the fundamental theorem of arithmetic*" or "the unique factorization theorem."

THEOREM 1.7 (*The Unique Factorization Theorem*). *If $n > 1$ is an integer, then n is a unique (up to a rearrangement of the factors) product*

$$n = p_1^{\alpha_1} \cdots p_r^{\alpha_r} \tag{1.1}$$

of distinct primes $p_1, \ldots, p_r$ with exponents α_j as positive integers.

PROOF. In view of Theorem 1.5, we have only to prove the uniqueness. If the theorem is false, we choose the smallest integer $n > 1$ that can be written in two ways as product of primes, say

$$n = p_1 \cdots p_r = q_1 \cdots q_s. \tag{1.2}$$

Clearly $p_i \neq q_j$ for each pair i, j, because otherwise by canceling a prime factor common to both sides of (1.2), we get an integer less than n with the required property. We may assume that $p_1 > q_1$. Put

$$m = (p_1 - q_1)p_2 \cdots p_r = q_1(q_2 \cdots q_s - p_2 \cdots p_r). \tag{1.3}$$

Now q_1 cannot appear in the factorization of $p_1 - q_1$, because otherwise by Theorem 1.2, part (6), $q_1 | p_1$. Thus (1.3) shows that m has two distinct factorizations. Because $m < n$, this contradicts the minimality of n. □

REMARK 1.8. In the expression (1.1), it is sometimes (cf. Theorem 1.14) useful to allow $\alpha_j = 0$.

COROLLARY 1.9. *If a prime $p | ab$, then either $p | a$ or $p | b$.*

PROOF. The prime p must appear in the factorization of either a or b. □

EXERCISE 1.10.

1. Suppose n is as in (1.1). Then n is a square, i.e., $n = m^2$ for some $m \in \mathbb{Z}$, if and only if α_j is even for all $j = 1, \ldots, r$.

2. If n is as in (1.1), it is called *square-free* if $\alpha_j = 1$ for all $j = 1, \ldots, r$. If n is square-free show that $\sqrt{n}$ is *irrational*, i.e., $\sqrt{n}$ cannot be written as

$$\sqrt{n} = \frac{a}{b}$$

with $a, b \in \mathbb{Z}$ and $b \neq 0$.

THEOREM 1.11 (*Euclid*). *There are infinitely many primes.*

PROOF. Suppose there are only finitely many primes $p_1, \ldots, p_r$. Put $N = p_1 \cdots p_r + 1 > 1$. By the unique factorization theorem,

$$N = p_1 \cdots p_r + 1 = \prod_{j=1}^{r} p_j^{\alpha_j} \qquad (\alpha_j \geq 0). \tag{1.4}$$

Because $N > 1$, there is at least one j such that $\alpha_j > 0$ and it follows from (1.4) that $p_j \mid 1$. This contradiction proves the theorem. □

EXERCISE 1.12. Show that there are arbitrarily large gaps between primes, i.e., given an integer $N \geq 1$, there is a pair of primes p_1, p_2 with $p_1 < p_2$ such that

1. $p_2 - p_1 \geq N$ and
2. There is no prime p with $p_1 < p < p_2$.

[*Hint.* None of the N consecutive integers

$$(N+1)! + j, \qquad j = 2, 3, \ldots, N+1$$

is a prime.]

DEFINITION 1.13. Let $a, b \in \mathbb{Z}$, both $a, b \neq 0$. Then the largest positive integer $d = (a, b)$ that divides both a and b is called the *greatest common divisor*, or the g.c.d., of a and b.

The smallest positive integer $l = [a, b]$ divisible by both a and b is called the *least common multiple*, or the l.c.m., of a and b.

If $a \neq b = 0$, we may define $(a, b) = |a|$, where $|x|$ denotes the absolute value of a real number x. Since 1 is a common divisor and $a \cdot b$ is a common multiple of a and b, it is obvious that (a, b) and $[a, b]$ are uniquely defined. Moreover, $(a, b) = (b, a) = (-a, b)$ and $[a, b] = [b, a] = [-a, b]$. The following theorem characterizes (a, b) and $[a, b]$.

THEOREM 1.14. *Suppose a, b are positive integers:*

$$a = \prod_{j=1}^{r} p_j^{\alpha_j} \qquad (\alpha_j \geq 0)$$

and

$$b = \prod_{j=1}^{r} p_j^{\beta_j} \qquad (\beta_j \geq 0).$$

Then

$$(a, b) = \prod_{j=1}^{r} p_j^{\min(\alpha_j, \beta_j)}.$$

$$[a, b] = \prod_{j=1}^{r} p_j^{\max(\alpha_j, \beta_j)}.$$

PROOF. The proof is obvious. □

DEFINITION 1.15. Two nonzero integers a and b are said to be *relatively prime*, or *coprime*, if $(a, b) = 1$.

THEOREM 1.16. *Suppose a, b, c, and m are all in* $\mathbb{Z}$ *and* $m \geq 1$. *Then*

1. $(ma, mb) = m(a, b)$ *and* $[ma, mb] = m[a, b]$.
2. *If* $d = (a, b)$ *then* $(a/d, b/d) = 1$.
3. *If* $a \mid b$, *then* $(a, b) = |a|$, $[a, b] = |b|$.
4. *If* $(a, b) = 1$, *then* $(a + mb, b) = 1$.
5. *If* $(a, m) = (b, m) = 1$, *then* $(ab, m) = 1$.
6. *If* $a \mid bc$ *and* $(a, b) = 1$, *then* $a \mid c$.
7. *For a prime p*, $(a, p) = 1$ *if and only if p does not divide a.*

PROOF. The proof is left as an exercise for the reader.

EXERCISE 1.17. Show by an example that $(a, b) = 1$ and $(b, c) = 1$ does not imply that $(a, c) = 1$.

THEOREM 1.18. *Suppose that both a and* $b > 0$ *and* $(a, b) = 1$. *Then ab is a square if and only if both a and b are squares.*

PROOF. The proof is left as an exercise for the reader.

THEOREM 1.19. *If* $(a, b) = d$, *there exist integers* λ *and* μ *such that* $\lambda a + \mu b = d$.

PROOF. First we observe that (a, b) is the positive integer d characterized by the following two conditions:

1. $d \mid a$ and $d \mid b$.
2. If $c \mid a$ and $c \mid b$, then $c \mid d$.

Now as x, y range over all the integers (positive, negative, or zero), $ax + by$ takes values that are positive, negative, and zero. Let $x = \lambda$, $y = \mu$ be such that $d = \lambda a + \mu b$ is the smallest positive value of $ax + by$. We show that

$d = (a, b)$. The condition (2) is obvious. To prove (1) first we show that for any x, y in $\mathbb{Z}$, $d \mid ax + by$. By division algorithm,

$$ax + by = qd + r, \qquad 0 \le r < d.$$

We must show that $r = 0$. But this is obvious, because otherwise

$$r = a(x - q\lambda) + b(y - q\mu)$$

would contradict the minimality of d. Now by taking $x = 1$, $y = 0$, it follows that $d \mid a$. Similarly, $d \mid b$. This proves (1), which completes the proof. □

COROLLARY 1.20. *If a, b are coprime, there exist integers λ and μ such that $\lambda a + \mu b = 1$.*

COROLLARY 1.21. *The linear equation*

$$ax + by = c$$

with a, b in $\mathbb{Z}$ has a solution in integers if and only if $(a, b) \mid c$.

Neither Theorem 1.19 nor its proof tells us how to find λ and μ or how to compute (a, b). The following theorem answers this question. If one of a or $b = 0$, the problem is trivial. Without loss of generality we may also assume that $a > b > 0$ and $b \nmid a$.

THEOREM 1.22 (*The Euclidean Algorithm*). *Suppose a and b are two integers with $a > b > 0$ and $b \nmid a$. By repeated application of the division algorithm, we write*

$$\left.\begin{aligned} a &= q_1 b + r_1, & 0 &< r_1 < b; \\ b &= q_2 r_1 + r_2, & 0 &< r_2 < r_1; \\ r_1 &= q_3 r_2 + r_3, & 0 &< r_3 < r_2; \\ &\vdots \\ r_{j-2} &= q_j r_{j-1} + r_j, & 0 &< r_j < r_{j-1} \end{aligned}\right\} \tag{1.5}$$

and

$$r_{j-1} = q_{j+1} r_j. \tag{1.6}$$

Then $(a, b) = r_j$, the last nonzero remainder in this process.

PROOF. It suffices to show that

1. $r_j \mid a$ and $r_j \mid b$.
2. If $d \mid a$ and $d \mid b$, then $d \mid r_j$.

But (1) [respectively, (2)] follows at once by running up (respectively, down) the chain of equations in (1.5) and (1.6). □

REMARK 1.23. The integers λ, μ can be computed by eliminating $r_1, r_2, \ldots, r_{j-1}$ from (1.5) as shown in the Example below.

EXAMPLE. If $a = 243$, $b = 198$, by division algorithm, we have

$$\left.\begin{aligned} 243 &= 1 \cdot 198 + 45; \\ 198 &= 4 \cdot 45 + 18; \\ 45 &= 2 \cdot 18 + 9; \end{aligned}\right\} \qquad (*)$$

$$18 = 2 \cdot 9.$$

Thus $(a, b) = 9$.

Now from $(*)$, we get

$$\begin{aligned} 9 &= 45 - 2 \cdot 18 \\ &= 45 - 2(198 - 4 \cdot 45) \\ &= 9 \cdot 45 - 2 \cdot 198 \\ &= 9(243 - 198) - 2 \cdot 198 \\ &= 9 \cdot 243 - 11 \cdot 198. \end{aligned}$$

Hence $\lambda = 9$, $\mu = -11$.

EXERCISE 1.24. If $a = 963$, $b = 657$, compute (a, b) and find λ, μ such that $(a, b) = \lambda a + \mu b$.

EXERCISE 1.25. Generalize Definition 1.13, Theorem 1.14, Definition 1.15, Theorem 1.16 parts (1), (2), and (5), Theorem 1.18, Theorem 1.19, Corollary 1.20, and Corollary 1.21 to the g.c.d. $(a_1, \ldots, a_n)$ and the l.c.m. $[a_1, \ldots, a_n]$ of $a_1, \ldots, a_n$.

1.4. Congruences

One of the most useful concepts in number theory, first introduced by Gauss, is that of congruence.

DEFINITION 1.26. Let $m \neq 0$ be a fixed integer, called the *modulus*, and let $a, b \in \mathbb{Z}$. We say that a is *congruent to b modulo m*, written as $a \equiv b \pmod{m}$, if $m \mid a - b$.

Thus, $a \equiv b \pmod{m}$ if a and b leave the same remainder on division by m. It is obvious from the definition that

1. $a \equiv a \pmod{m}$;
2. If $a \equiv b \pmod{m}$, then $b \equiv a \pmod{m}$;
3. If $a \equiv b \pmod{m}$ and $b \equiv c \pmod{m}$, then $a \equiv c \pmod{m}$.

Thus the congruence is an "*equivalence relation.*" Also it is immediate from the definition that if $a \equiv b \pmod{m}$ and $n \mid m$, then $a \equiv b \pmod{n}$.

THEOREM 1.27. *Suppose* $x_i \equiv y_i \pmod{m}$, $i = 1, 2$, *and* $c \in \mathbb{Z}$. *Then*

1. $x_1 + x_2 \equiv y_1 + y_2 \pmod{m}$;
2. $x_1 x_2 \equiv y_1 y_2 \pmod{m}$;
3. $cx_1 \equiv cy_1 \pmod{m}$.

PROOF. The proof is left as an exercise for the reader.

COROLLARY 1.28. *Suppose* $f(x) \in \mathbb{Z}[x]$, i.e., $f(x)$ *is a polynomial with integer coefficients and* $a \equiv b \pmod{m}$. *Then* $f(a) \equiv f(b) \pmod{m}$.

EXERCISE 1.29. Suppose a, b are odd and c is even. Show that

1. $a^2 \equiv 1 \pmod{8}$, in particular, $a^2 \equiv 1 \pmod{4}$,
2. $a^2 + b^2 \equiv 2 \pmod{4}$,
3. $c^2 \equiv 0 \pmod{4}$,
4. $a^2 + c^2 \equiv 1 \pmod{4}$.

THEOREM 1.30. *If* $(a, m) = 1$, *then*

$$ax \equiv 1 \pmod{m}$$

has a solution.

PROOF. Since $(a, m) = 1$, by Corollary 1.20, there are integers λ, μ such that

$$\lambda a + \mu m = 1$$

so $a\lambda \equiv 1 \pmod{m}$. □

THEOREM 1.31 (*Chinese Remainder Theorem*). *Given r nonzero integers* $m_1, \ldots, m_r$ *which are coprime in pairs* [i.e., $(m_i, m_j) = 1$ *if* $i \neq j$] *and r arbitrary integers* $a_1, \ldots, a_r$, *the congruences*

$$x \equiv a_j \pmod{m_j}, \qquad j = 1, \ldots, r$$

have a common solution. Moreover, if x and y are two common solutions, then

$$x \equiv y \pmod{m},$$

where $m = m_1 \cdots m_r$.

PROOF. Because $(m/m_j, m_j) = 1$, by Theorem 1.30, there are integers $x_1, \ldots, x_r$ such that

$$(m/m_j)x_j \equiv 1 \pmod{m_j}.$$

Clearly, if $i \neq j$ we have $(m/m_j)x_j \equiv 0 \pmod{m_i}$. Therefore we have

$$\begin{aligned} x &\stackrel{\text{def}}{=} \sum_{j=1}^{r} \frac{m}{m_j} a_j x_j \\ &\equiv \frac{m}{m_i} x_i a_i \\ &\equiv a_i \pmod{m_i}. \end{aligned}$$

The last statement follows at once from the observation that if $a \mid c$, $b \mid c$, and $(a, b) = 1$, then $ab \mid c$. □

1.5 Diophantine Equations

A major goal of this book is the study of *diophantine equations*, which are polynomial equations

$$f_j(x_1, \ldots, x_n) = 0, \qquad j = 1, \ldots, m \tag{1.7}$$

with $f_j(x_1, \ldots, x_n)$ in $\mathbb{Z}[x_1, \ldots, x_n]$. We assume that $n \geq 2$. There are two problems: To find (1) integer solutions and (2) rational solutions of (1.7). As examples we list some of the best-known diophantine problems, which have been studied for hundreds, or even thousands, of years:

(1) Pythagorean Triangles. To find right triangles whose sides are whole numbers; i.e., to solve the equation

$$x^2 + y^2 = z^2 \tag{1.8}$$

in integers with all $x, y, z \neq 0$.

(2) Fibonacci Curve. The following simultaneous equations were first studied (in 1220) by Leonardo Pissano, commonly known as Fibonacci:

$$\begin{aligned} x^2 + y^2 &= z^2, \\ x^2 - y^2 &= t^2. \end{aligned} \tag{1.9}$$

(3) Fermat's Equation. The equation (1.8) may be generalized to

$$x^n + y^n = z^n \qquad (n \in \mathbb{N}, n \geq 3). \tag{1.10}$$

All of these are "homogeneous" equations. If all the equations in (1.7) are homogeneous, there is always the *trivial solution*, viz., $x_1 = \cdots = x_n = 0$. Moreover, it is obvious that in this case the problems (1) and (2) above are equivalent. However, if the equations are not homogeneous, for example,

$$y^2 = x^3 - 17x,$$

in general the problems (1) and (2) have to be treated separately.

In the rest of this chapter we shall study the first two examples. Fermat's equation is still far from solved. There is a conjecture (which Fermat claimed to have proved, but the margin of his note book was too small for the proof) known as Fermat's last theorem.

FERMAT'S LAST THEOREM. *For all* $n \geq 3$, *the equation*

$$x^n + y^n = z^n$$

has no solution in integers with $xyz \neq 0$.

This has been proved for certain values of n. We shall give an elementary proof for $n = 4$ (cf. Corollary 1.36). For $n = 3$, however, the proof is not as elementary as for $n = 4$ (cf. Theorem 3.16 in Ref. 1).

For the rest of this section we shall discuss equation (1.8). Any solution (x, y, z) of (1.8) is called *nontrivial* if $xyz \neq 0$. Some of the well-known nontrivial solutions of (1.8) are $(3, 4, 5)$ and $(5, 12, 13)$, while some others like $(6, 8, 10)$ and $(15, 36, 39)$ are obtained as (cx, cy, cz) from (x, y, z), for an integer $c \neq 0$. Any two solutions (x, y, z) and (cx, cy, cz) with $c \neq 0$ are *equivalent* and shall not be regarded as different.

If (x, y, z) is a nontrivial solution (in integers) of (1.8) and d is a common divisor of any two of x, y, z, it must divide the third as well. So $(x/d, y/d, z/d)$ is also an integer solution of (1.8). Thus it follows that among all the equivalent solutions, there is one, say (x, y, z), such that x, y, z are coprime in pairs and any solution in its *equivalence class* is (cx, cy, cz) for some $c \neq 0$. Such a solution (x, y, z) is called a *primitive solution* of (1.8), if all $x, y, z > 0$. Any solution of (1.8) is obtained from a primitive solution (x, y, z) as (cx, cy, cz) for an integer c and if necessary by changing some of the signs of x, y, z. We may also switch the role of x and y without regarding it as a new solution. Thus to determine all the nontrivial solutions of (1.8), it suffices to prove the following theorem.

THEOREM 1.32. *Suppose* (x, y, z) *is a primitive solution of* (1.8). *Then one of* x, y *is even and the other odd. If* x *is odd, the solution is of the form*

$$x = a^2 - b^2, \qquad y = 2ab, \qquad z = a^2 + b^2 \tag{1.11}$$

where a, b *are integers satisfying*

$$\begin{aligned} a - b &\equiv 1 \pmod 2, \\ a > b > 0 \quad &\text{and} \quad (a, b) = 1. \end{aligned} \tag{1.12}$$

Conversely, any triplet (x, y, z) *given by* (1.11) *and* (1.12) *is a primitive solution of* (1.8).

PROOF. Because (x, y, z) is a primitive solution, x and y, cannot both be even. They cannot both be odd either, because then $x^2 + y^2 \equiv 2 \pmod 4$, whereas $z^2 \equiv 0$ or $1 \pmod 4$. We assume that x is odd and y is even. Rewrite (1.8) as

$$\left(\frac{z+x}{2}\right)\left(\frac{z-x}{2}\right) = \left(\frac{y}{2}\right)^2. \tag{1.13}$$

Note that all three terms are integers. If

$$d = \left(\frac{z+x}{2}, \frac{z-x}{2}\right),$$

then

$$d \mid x = \frac{z+x}{2} - \frac{z-x}{2}$$

and

$$d \mid z = \frac{z+x}{2} + \frac{z-x}{2}.$$

Since $(x, z) = 1$, we must have $d = 1$ and by Theorem 1.18, we can write

$$\frac{z+x}{2} = a^2 \quad \text{and} \quad \frac{z-x}{2} = b^2 \tag{1.14}$$

with $a > b > 0$. Clearly, $z = a^2 + b^2$, $x = a^2 - b^2$ and by (1.13), $y = 2ab$. Since $d = 1$, it is obvious from (1.14) that $(a, b) = 1$. It is also obvious that $a - b \equiv 1 \pmod 2$, because otherwise a, b are of the same *parity*, i.e., they are either both odd or both even, in each case implying that x, z are both even, which contradicts the primitivity of (x, y, z).

Conversely, if a, b are as in (1.12) and x, y, z as in (1.11), then (x, y, z) is a solution of (1.8), all $x, y, z > 0$ and x, z are both odd. If $d = (x, z)$, then $d \mid 2a^2$ and $d \mid 2b^2$. Since d is odd, $d \mid a^2$ and $d \mid b^2$. Now $(a, b) = 1$ implies that $d = 1$, proving that (x, y, z) is a primitive solution. □

1.6. Congruent Numbers

DEFINITION 1.33. A positive integer A is called a *congruent number* if it is the area of a right triangle of rational sides, i.e., if

$$A = \frac{ab}{2},$$

and (1.15)

$$a^2 + b^2 = h^2 \qquad (a, b, h \in \mathbb{Q}).$$

Congruent numbers should not be confused with the congruences of Section 1.4 of this chapter. Some examples of congruent numbers are 6 and 30, which are the areas of the Pythagorean triangles with sides $(3, 4, 5)$ and $(5, 12, 13)$, respectively. The smallest congruent number discovered in 1220 by Leonard Pissano is 5. It is the area of the right triangle of rational sides $3/2, 20/3, 41/6$. Suppose $c \in \mathbb{N}$. Then A is a congruent number if and only if c^2A is a congruent number. Thus without loss of generality, we may assume that A is square-free.

THEOREM 1.34. *A positive integer A is a congruent number if and only if the simultaneous equations*

$$\begin{aligned} x^2 + Ay^2 &= z^2 \\ x^2 - Ay^2 &= t^2 \end{aligned} \tag{1.16}$$

have a solution in integers with $y \neq 0$.

Such a solution is called *a nontrivial solution.*

PROOF. Recall that since (1.16) is homogeneous a nontrivial solution in integers is equivalent to a nontrivial solution in rationals. If A is a congruent number, by (1.15),

$$b = \frac{2A}{a}$$

and therefore,

$$\begin{aligned} h^2 &= a^2 + b^2 \\ &= a^2 + \frac{4A^2}{a^2}, \end{aligned}$$

which gives

$$\left(\frac{h}{2}\right)^2 = \left(\frac{a}{2}\right)^2 + \left(\frac{A}{a}\right)^2.$$

Adding $\pm A$ to each side of the above equation to complete a square on the right, we obtain

$$\left(\frac{h}{2}\right)^2 \pm A = \left(\frac{a}{2} \pm \frac{A}{a}\right)^2.$$

Now put $x = h/2$, $y = 1$, $z = a/2 + A/a$, and $t = a/2 - A/a$ and we have a nontrivial solution of (1.16) in rationals and hence in integers.

Conversely, a nontrivial solution leads to a rational solution $(x, 1, z, t)$ of (1.16), i.e.,

$$\begin{aligned} x^2 + A &= z^2, \\ x^2 - A &= t^2. \end{aligned} \tag{1.17}$$

From (1.17), $2A = z^2 - t^2$ or

$$A = \frac{(z+t)(z-t)}{2}$$

and $2x^2 = z^2 + t^2$, showing that $(z + t, z - t, 2x)$ are the sides of a right triangle of area A. □

THEOREM 1.35 (*Leonardo Pissano, 1220*). *1 is not a congruent number.*

PROOF (*Fermat*). We shall show that there is no nontrivial solution in integers of

$$\begin{aligned} x^2 + y^2 &= z^2, \\ x^2 - y^2 &= t^2. \end{aligned} \tag{1.18}$$

Suppose there is one. We can assume that x, y, z, t are coprime in pairs. Since $y \neq 0$, it is obvious that none of x, z, t is zero either. Therefore, there is no loss of generality in assuming that $x, y, z \geq 1$ and $t \neq 0$.

Among all such nontrivial solutions we choose (x, y, z, t) with the smallest $y \geq 1$. By a method known as *Fermat descent* we manipulate (x, y, z, t) to obtain a similar solution (x_1, y_1, z_1, t_1) with $y > y_1 \geq 1$ to contradict the choice of (x, y, z, t).

First we note that y cannot be odd; otherwise subtracting the second equation in (1.18) from the first, we obtain

$$2y^2 = z^2 - t^2 \equiv 0 \quad \text{or} \quad \pm 1 \pmod 4$$

while

$$2y^2 \equiv 2 \pmod 4,$$

which is impossible. Now since y is even, x, z, t are all odd.

From equations (1.18) we obtain

$$\begin{aligned} y^2 &= (z+x)(z-x), \\ y^2 &= (x+t)(x-t), \end{aligned} \tag{1.19}$$

$$2y^2 = (z+t)(z-t). \tag{1.20}$$

Also by adding the equations in (1.18) and multiplying by 2, we have

$$\begin{aligned} 4x^2 &= 2(z^2 + t^2) \\ &= (z+t)^2 + (z-t)^2, \end{aligned}$$

which gives

$$(z-t)^2 = (2x+z+t)(2x-z-t). \tag{1.21}$$

The two factors on the right of any equation in (1.19), (1.20), and (1.21) have no odd prime factor in common; otherwise if, for example, $p \mid z+x$ and $p \mid z-x$, then $p \mid z$ and $p \mid x$, implying that x and z are not coprime. To take care of (1.21), use the fact that $p \mid z+t$ implies $p \mid y$, by (1.20).

If we can show that there are positive integers x_1, y_1, z_1, t_1, satisfying

$$\begin{aligned} z - x &= 2y_1^2, \\ x - t &= 2x_1^2, \\ 2x - z - t &= 2t_1^2, \end{aligned} \tag{1.22}$$

and

$$z - t = 2z_1^2, \tag{1.23}$$

by adding and subtracting the first two equations in (1.22) and then using (1.23) and the last equation in (1.22), we see that (x_1, y_1, z_1, t_1) is a solution of (1.18). Further, $2y_1^2 = z - x \mid (z-x)(z+x) = y^2$ would imply that $y > y_1 \geq 1$, contradicting the minimality of y.

The exact power of 2 dividing each factor on the right of (1.19) and (1.21) is one, because if, for example

$$z + x = 2^\alpha a, \qquad z - x = 2^\beta b, \qquad y = 2^\gamma c$$

with $\alpha, \beta, \gamma \geq 1$ and a, b, c all odd, then from (1.19), $\alpha + \beta = 2\gamma$. If $\alpha > 1$, then $\beta, \gamma > 1$, implying that 2 is a common factor of z and x, a contradiction. So we can rewrite (1.19) and (1.21) as

$$\left(\frac{y}{2}\right)^2 = \left(\frac{z+x}{2}\right)\left(\frac{z-x}{2}\right),$$

$$\left(\frac{y}{2}\right)^2 = \left(\frac{x+t}{2}\right)\left(\frac{x-t}{2}\right),$$

$$\left(\frac{z-t}{2}\right)^2 = \left(\frac{2x+z+t}{2}\right)\left(\frac{2x-z-t}{2}\right),$$

where the two factors on the right of each of these equations are coprime. Thus each of these factors is a square and we get (1.22).

In (1.20) by changing the sign of t if necessary, it is easy to see that

$$\begin{aligned} z + t &= 4r, \\ z - t &= 2s, \end{aligned}$$

with r, s odd. Therefore, (1.20) can be rewritten as

$$\left(\frac{y}{2}\right)^2 = \left(\frac{z+t}{4}\right)\left(\frac{z-t}{2}\right)$$

again with coprime factors on the right. This gives us (1.23). □

COROLLARY 1.36. *The equation*

$$x^4 + y^4 = z^4 \tag{1.24}$$

has no solution in integers with all x, y, $z \neq 0$.

PROOF. If there is such a solution we may choose one with all x, y, z coprime in pairs. Then y is even and x, z are odd. If we rewrite (1.24) as

$$(z^2 + y^2)(z^2 - y^2) = x^4, \tag{1.25}$$

we see that the product of two (odd and) mutually coprime positive integers on the left of (1.25) is a square. Therefore, by Theorem 1.18, we get a nontrivial solution

$$z^2 + y^2 = u^2$$
$$z^2 - y^2 = v^2$$

of equations (1.18), which is a contradiction. □

REMARK 1.37. We have seen that the existence of a nontrivial solution of (1.16) is equivalent to A being a congruent number. We will say more about congruent numbers later on in this book. More generally, we can consider the equations

$$x^2 + My^2 = z^2$$
$$x^2 + Ny^2 = t^2$$

where M and N are two nonzero square-free integers. Such equations have been extensively studied by Ono. For details see Ref. 2.

References

1. W. J. LeVeque, *Topics in Number Theory*, Vol. II, Addison Wesley, Reading, Massachusetts (1956).
2. T. Ono, *Variation on a Theme of Euler*, Jikkyō, Tokyo (1980) (in Japanese).

2

Algebraic Methods

There are concepts in number theory that are best expressed in the language of algebra. We shall discuss algebra only to the extent needed for our purpose.

2.1. Groups

DEFINITION 2.1. A *group* is a pair $(G, *)$ consisting of a nonempty set G and a binary operation $*$ (which assigns to each ordered pair of elements x, y of G a unique element $x * y$ of G) such that

1. $(x * y) * z = x * (y * z)$ for all x, y, z in G;
2. There is an element e in G, called the *identity*, such that

$$e * x = x * e = x \text{ for all } x \text{ in } G;$$

3. For each x in G there is an element y of G, called the *inverse* of x and written as x^{-1}, such that

$$x * y = y * x = e.$$

Furthermore, $(G, *)$ is called *abelian* if $x * y = y * x$ for all x, y in G.

REMARKS 2.2.
1. The identity e is unique.
2. Each element has a unique inverse.

EXAMPLES 2.3.
1. Let $G = \mathbb{Z}, \mathbb{Q}, \mathbb{R}$, or $\mathbb{C}$. Then $(G, +)$ is an abelian group.
2. Let $A = \mathbb{Z}, \mathbb{Q}, \mathbb{R}$, or $\mathbb{C}$ and $A[x]$ consist of all polynomials $f(x)$ with coefficients in A. Then $(A[x], +)$ is an abelian group.
3. Let $k = \mathbb{Q}, \mathbb{R}$, or $\mathbb{C}$ and $k^\times = \{x \in k \mid x \neq 0\}$. Then $k^\times$ is a group under multiplication.

4. Let $A = \mathbb{Z}, \mathbb{Q}, \mathbb{R}$, or $\mathbb{C}$. The set $M(n, A)$ of $n \times n$ matrices $x = (x_{ij})$ with x_{ij} in A is a group under addition. We let $\mathrm{GL}(n, A)$, called the *general linear group of $n \times n$ matrices over A*, denote the set of all invertible matrices in $M(n, A)$ whose inverse also has entries in A. Then for $n \geq 2$, $\mathrm{GL}(n, A)$ is a non-abelian group under the operation of multiplication of matrices. In particular,
$$\mathrm{GL}(n, \mathbb{Z}) = \{x \in M(n, \mathbb{Z}) \mid \det(x) = \pm 1\}.$$

EXERCISE 2.4. Suppose $d > 1$ is a square-free integer. Let G consist of all solutions (x, y) of

$$x^2 - dy^2 = 1 \tag{2.1}$$

with x, y in $\mathbb{Z}$. Given two solutions (x_i, y_i), $i = 1, 2$, define

$$\begin{aligned}(X, Y) &= (x_1, y_1) * (x_2, y_2)\\ &= (x_1x_2 + dy_1y_2, x_1y_2 + x_2y_1).\end{aligned}$$

Show that

1. (X, Y) is a solution of (2.1);
2. $(G, *)$ is an abelian group with $(1, 0)$ as the identity and $(x, -y)$ as the inverse of (x, y).

DEFINITION 2.5. A group $(G, *)$ is *finite* or *infinite* according as the set G has only finitely many elements or not. If $(G, *)$ is finite, its *order*, denoted as $|G|$ or $\mathrm{ord}(G)$, is the number of elements in G. We say that a group $(G, *)$ is *of finite* or *infinite order* according as $|G|$ is finite or not.

All the examples given so far are of infinite groups. (That the group G in Exercise 2.4 is an infinite group is a nontrivial fact; cf. Chapter 4.)

Let m be any integer and

$$m\mathbb{Z} = \{mx \mid x \in \mathbb{Z}\},$$

i.e., $m\mathbb{Z}$ consists of all the integer multiples of m. Then $(m\mathbb{Z}, +)$ is another infinite group. We now give some examples of finite groups.

EXAMPLE 2.6. Let m be a positive integer and

$$R_m = \{0, 1, \ldots, m-1\}$$

be the set of all possible remainders $r (0 \leq r < m)$ of integers on division by m. (R_m is often denoted by $\mathbb{Z}_m$, which is confusing when $m = p$.) For r, s in R_m let $r + s$ also denote the remainder t $(0 \leq t < m)$ of the usual sum $r + s$ on division by m. Then $(R_m, +)$ is a finite abelian group of order m. The identity is 0 and the inverse of $r \neq 0$ is $m - r$. (Identity is always its own inverse.)

EXAMPLE 2.7. For an integer $m \geq 1$, let

$$\zeta = \zeta_m \stackrel{\text{def}}{=} \exp\left(\frac{2\pi\sqrt{-1}}{m}\right) = \cos\frac{2\pi}{m} + \sqrt{-1}\sin\frac{2\pi}{m}.$$

Using *De Moivre's theorem*, which states that

$$(\cos\theta + \sqrt{-1}\sin\theta)^n = \cos n\theta + \sqrt{-1}\cos n\theta$$

for any integer n, it is easy to see that

$$\mu_m = \{\zeta^n \mid n \in \mathbb{Z}\}$$

is a finite group of order m under the multiplication of complex numbers. We call it the group of mth *roots of unity*, because μ_m consists of all the roots of the equation

$$x^m = 1.$$

In fact, $\mu_m = \{\zeta^n \mid n = 0, 1, \ldots, m-1\}$.

From now on we shall write G for $(G, *)$ and xy for $x * y$.

DEFINITION 2.8. Suppose

$$f: X \to Y$$

is a function. Then

1. f is *one-to-one* or *injective* if $f(x_1) = f(x_2)$ is possible only if $x_1 = x_2$;
2. f is *onto* or *surjective* if for each y in Y there is an x in X with $y = f(x)$;
3. f is *bijective* if f is injective as well as surjective.

EXERCISE 2.9. Give examples of functions

$$f: X \to Y$$

such that

1. f is neither injective nor surjective;
2. f is injective, but not surjective;
3. f is surjective, but not injective;
4. f is injective and surjective, i.e., bijective.

THEOREM 2.10. *Suppose G is a group and $x \in G$. Then the function*

$$m_x: G \to G$$

defined by $m_x(y) = xy$ is bijective. In particular, if G is finite, m_x permutes the elements of G among themselves.

PROOF. If $m_x(y_1) = m_x(y_2)$, then $xy_1 = xy_2 \Rightarrow x^{-1}(xy_1) = x^{-1}(xy_2) \Rightarrow y_1 = y_2$. Hence m_x is injective. If $y \in G$, then $m_x(x^{-1}y) = y$. So m_x is surjective. □

2.2. Subgroups

DEFINITION 2.11. A non-empty subset H of a group G is called a *subgroup* of G if $x^{-1}y \in H$ for all x, y in H.

EXAMPLES 2.12.
1. $m\mathbb{Z}$ is a subgroup of $\mathbb{Z}$.
2. μ_m is a subgroup of $\mathbb{C}^\times$.
3. Put

$$\mathrm{SL}(n, A) = \{x \in \mathrm{GL}(n, A) \mid \det(x) = 1\}.$$

Then $\mathrm{SL}(n, A)$ is a subgroup of $\mathrm{GL}(n, A)$. We call $\mathrm{SL}(n, A)$ the *special linear group of $n \times n$ matrices over A.*

REMARK 2.13. It is clear that $e \in H$, for H being nonempty contains an element x and $e = x^{-1}x$.

EXERCISES 2.14. Suppose H_1, H_2 are two subgroups of a group G. Show the following:

1. $H_1 \cap H_2$ is a subgroup of G.
2. $H_1 \cup H_2$ need not be a subgroup of G.
3. If G is an abelian group written additively and $n > 1$ is an integer, show that $nG = \{nx \mid x \in G\}$ is a subgroup of G.

THEOREM 2.15 (*Lagrange*). *If H is a subgroup of a finite group G, then* $\mathrm{ord}(H) \mid \mathrm{ord}(G)$.

PROOF. Let

$$H = \{h_1, \ldots, h_r\}.$$

If $H = G$, there is nothing to prove. Otherwise, there is an element g_1 in G that is not in H. If

$$g_1H = \{g_1h_j \mid j = 1, \ldots, r\},$$

then $H \cap g_1H = \varnothing$, because otherwise $g_1h_i = h_j$ for some i and j, which implies that

$$g_1 = h_jh_i^{-1} \in H.$$

Now either $G = H \cup g_1 H$ or this process can be continued until we get

$$G = H \cup g_1 H \cup \cdots \cup g_{s-1} H$$

as a disjoint union [X is called a *disjoint union* of X_1, X_2 if (1) $X = X_1 \cup X_2$ and (2) $X_1 \cap X_2 = \emptyset$] of $H, g_1 H, \ldots, g_{s-1} H$. Because each $g_j H$ has r elements, $\operatorname{ord}(G) = rs$. □

COROLLARY 2.16. *If G is a group of prime order, then it has only two subgroups G and $\{e\}$ called the trivial subgroups of G.*

2.3. Quotient Groups

Let H be a subgroup of a group G. For an element x of G, the set

$$xH = \{xh \mid h \in H\}$$

is called a *(left) coset* of H in G and x is called a *coset representative* of xH in G. Obviously $H = eH$ is a coset. It is easily seen (as in the proof of Theorem 2.15) that

1. any two cosets are either equal or disjoint;
2. $xH = yH$ if and only if $x^{-1}y \in H$.

Let

$$G/H = \{gH \mid g \in G\}$$

be the set of cosets of H in G.

DEFINITION 2.17. The number of cosets of H in G, i.e., the cardinality of the set G/H, if it is finite, is called the *index* of H in G and is denoted by $[G: H]$.

For example, $[\mathbb{Z}: m\mathbb{Z}] = m$ and $[G: \{e\}] = \operatorname{ord}(G)$.

EXERCISE 2.18. If K is a subgroup of H and H is a subgroup of G, show that K is a subgroup of G. If $[G: K]$ is finite, show that $[G: K] = [G: H][H: K]$.

THEOREM 2.19. *Let H be a subgroup of an abelian group G. Then the set G/H (read as G mod H) becomes a group called the quotient of G by H if we define the product of two cosets by*

$$(xH)(yH) = xyH.$$

PROOF. The proof is left as an exercise for the reader.

EXAMPLES 2.20.

1. We shall see (cf. Examples 2.31) that the quotient group $\mathbb{Z}/m\mathbb{Z}$ is "essentially the same" as R_m of Example 2.6.

2. Let G be $\mathbb{C}$ considered as a group under addition. Let ω_1, ω_2 be two nonzero complex numbers such that $\omega_1/\omega_2 = \tau$ is not a real number, i.e., ω_1 and ω_2 are linearly independent vectors in $\mathbb{C}$ considered as a vector space over $\mathbb{R}$. Then

$$L = \{m\omega_1 + n\omega_2 \mid m, n \in \mathbb{Z}\}$$

is a subgroup of G, called a *lattice* (see Fig. 2.1).

The quotient of an abelian group G (usually written additively) by a subgroup H is the group of equivalence classes of elements of G. Two elements α, β of G are considered to be identical or in the same equivalence class if they differ by an element of H, i.e., if $\alpha - \beta \in H$.

To see what $\mathbb{C}/L$ looks like, note that every complex number is in the equivalence class of one and only one $z = x\omega_1 + y\omega_2$ with x, y in $\mathbb{R}$ and $0 \leq x, y < 1$. In other words each coset of $\mathbb{C}/L$ has a unique coset representative z in the so-called *fundamental parallelogram*

$$T = \{z = x\omega_1 + y\omega_2 \mid x, y \in \mathbb{R} \text{ and } 0 \leq x, y < 1\}.$$

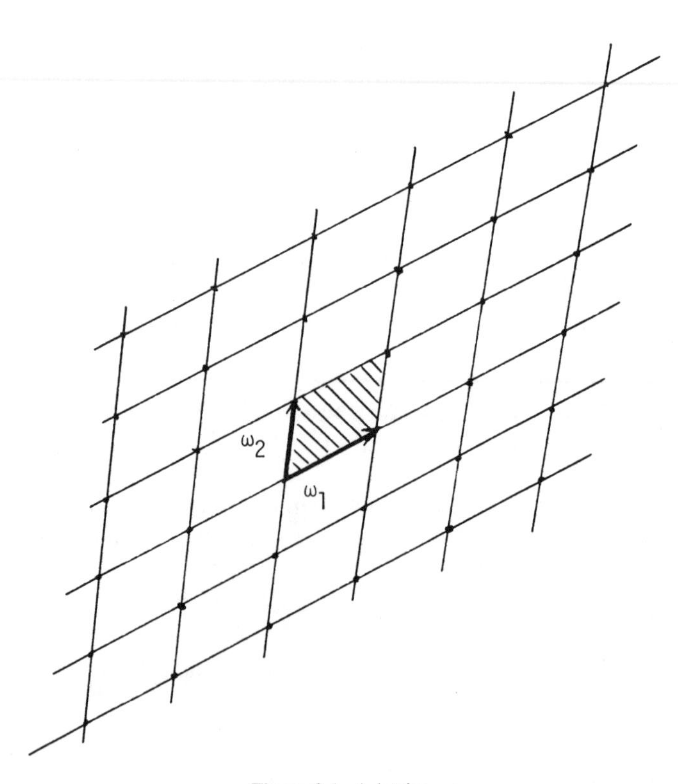

Figure 2.1. A lattice.

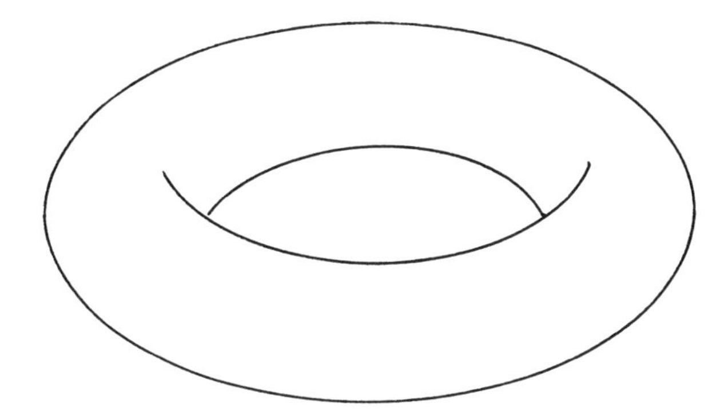

Figure 2.2. A torus.

So $\mathbb{C}/L$ can be identified with T, which consists of the opposite sides of

$$\{z = x\omega_1 + y\omega_2 \mid x, y \in \mathbb{R} \text{ and } 0 \leq x, y \leq 1\}$$

glued together. We first glue its horizontal sides to get a tube and then glue the ends of this tube. In this way we get a *torus* shown in Fig. 2.2.

2.4. Order of Elements

DEFINITION 2.21. Suppose x is an element of a group G. We say that x is of *finite order* if there is a positive integer d such that

$$x^d = \underbrace{x \cdots x}_{d \text{ times}} = e. \tag{2.2}$$

The least positive integer $d = \text{ord}(x)$ satisfying (2.2) is called the *order* of x. If no such d exists, x is said to be of *infinite order*.

The following are obvious:

1. *If m is a positive integer such that $x^m = e$, then $d \mid m$.* For otherwise, writing $m = qd + r$ $(0 < r \leq d - 1)$, we get

$$x^r = (x^d)^q x^r = x^{qd+r} = x^m = e,$$

contradicting the minimality of d.

2. *If G is a group of order m, then for all x in G, $\text{ord}(x) \mid \text{ord}(G)$. In particular,*

$$x^{\text{ord}(G)} = e,$$

for all x in G. This follows (by Theorem 2.15) from the following two facts

(i) Because G is finite, the powers

$$e = x^0, x = x^1, x^2, x^3, \ldots$$

of x must repeat, i.e., $x^i = x^j$ (for some $i > j$), which gives $x^{i-j} = e$. Thus all members of G are of finite order.

(ii) The set

$$H = \{e = x^0, x^1, x^2, \ldots, x^{d-1}\}$$

is a subgroup of G of order $d = \text{ord}(x)$.

EXERCISES 2.22.

1. Suppose an abelian group G has elements of order m and n. Show that G has an element of order $[m, n]$. *Hint*: Let x have order m and y have order n. Show (a) if $(m, n) = 1$, then xy has order mn; (b) if $m_0 | m$, then x^{m/m_0} has order m_0; (c) there exist $m_0 | n$ and $n_0 | n$ such that $(m_0, n_0) = 1$ and $m_0 n_0 = [m, n]$.

2. An element x of G is called a *torsion element* if x is of finite order. Suppose G is an abelian group. Put

$$G_{\text{tor}} = \{x \in G | x \text{ is a torsion element}\}.$$

Show that G_{tor} is a subgroup of G. It is called the *torsion subgroup* of G.

3. If G is an abelian group, show that G/G_{tor} is a *torsion-free group*, i.e., it has no torsion element $x \neq e$.

4. Suppose G is an abelian group and $N > 0$ is an integer. Put

$$G[N] = \{x \in G | \text{ord}(x) \text{ divides } N\}.$$

Show that $G[N]$ is a subgroup of G.

REMARK 2.23. If $\text{ord}(G)$ is finite, then $G_{\text{tor}} = G$. However, if $G_{\text{tor}} = G$, G need not be of finite order. For example

$$\mathbb{Q}^{\times 2} = \{x^2 | x \in \mathbb{Q}^{\times}\}$$

is a subgroup of $\mathbb{Q}^{\times}$. If we put

$$G = \mathbb{Q}^{\times}/\mathbb{Q}^{\times 2},$$

then G is an infinite group, but each element of G is of order 2 and hence $G_{\text{tor}} = G$.

2.5. Direct Product of Groups

Let G_1, G_2 be two groups. On the *Cartesian product* $G_1 \times G_2 = \{(g_1, g_2) | g_j \in G_j, j = 1, 2\}$ we define an operation by

$$(g_1, g_2)(g_1', g_2') = (g_1 g_1', g_2 g_2').$$

Then $G_1 \times G_2$ becomes a group called the *direct product* of G_1, G_2. If e_j denotes the identity of G_j, then (e_1, e_2) is the identity of $G_1 \times G_2$ and $(g_1, g_2)^{-1} = (g_1^{-1}, g_2^{-1})$. Obviously, (1) if G_1, G_2 are abelian, then so is $G_1 \times G_2$ (this is usually written as $G_1 \oplus G_2$ and called the *direct sum* of G_1, G_2); (2) if G_1, G_2 are finite groups, then so is $G_1 \times G_2$ and $|G_1 \times G_2| = |G_1||G_2|$.

EXERCISES 2.24.

1. Suppose H_j is a subgroup of $G_j, j = 1, 2$. Show that $H_1 \times H_2$ is a subgroup of $G_1 \times G_2$.
2. Suppose $G = \mathbb{Z}/p^n\mathbb{Z}$ ($n \geq 1$ is an integer and p is a prime). Show that

$$[G\colon 2G] = \begin{cases} 1, & \text{if } p > 2; \\ 2, & \text{if } p = 2. \end{cases}$$

3. Suppose G_1, G_2 are abelian groups, $G = G_1 \times G_2$ and the index $[G\colon 2G]$ is finite. Show that $[G\colon 2G] = [G_1\colon 2G_1][G_2\colon 2G_2]$.

NOTE. For an integer $n > 2$, the *direct product* $G_1 \times \cdots \times G_n$ and the *direct sum* $G_1 \oplus \cdots \oplus G_n$ are defined analogously.

2.6. Generators of a Group

An abelian group G is said to be *finitely generated* if G has finitely many elements $g_1, \ldots, g_r$ such that any g in G can be written as

$$g = g_1^{m_1} \cdots g_r^{m_r} \qquad (m_j \in \mathbb{Z}). \tag{2.3}$$

If the group operation is denoted additively, (2.3) takes the form

$$g = m_1 g_1 + \cdots + m_r g_r \qquad (m_j \in \mathbb{Z}).$$

We also say that G is *generated by* $g_1, \ldots, g_r$. If G is generated by one element, then it is called a *cyclic* group.

EXAMPLES 2.25.

1. $\mathbb{Z}$ is an infinite cyclic group generated by 1.
2. $\mathbb{Z}/m\mathbb{Z}$ (respectively, μ_m) is a finite cyclic group generated by $1 + m\mathbb{Z}$ (respectively, $\zeta = \cos(2\pi/m) + i \sin(2\pi/m)$).
3. $\mathbb{Z} \times \mathbb{Z} \times \mathbb{Z}$ is generated by three elements $\mathbf{e}_1 = (1, 0, 0)$, $\mathbf{e}_2 = (0, 1, 0)$, and $\mathbf{e}_3 = (0, 0, 1)$.
4. Any finite group is finitely generated.
5. If G_1, G_2 are finitely generated, then so is $G_1 \times G_2$.
6. $(\mathbb{R}, +)$, $(\mathbb{Q}^\times, \cdot)$, and $(\mathbb{Q}^\times/\mathbb{Q}^{\times 2}, \cdot)$ are not finitely generated.

EXERCISES 2.26.

1. Suppose G is a finitely generated abelian group and H is a subgroup of G. Show the following:

   i. G/H is finitely generated;
   ii. H is finitely generated. [This is not completely obvious. Try first the case in which G has one generator. Then use induction on the number of generators of G to treat the general case.]
   iii. G_{tor} is finite.

2. If a finite abelian group G has an element x with $\text{ord}(x) = \text{ord}(G)$, show that G is cyclic.

2.7. Homomorphisms of Groups

DEFINITION 2.27. Suppose G_1, G_2 are two groups. A map $f: G_1 \to G_2$ is called a (*group*) *homomorphism*, if $f(xy) = f(x)f(y)$ for all x, y in G_1.

EXAMPLES 2.28.

1. The determinant det: $\text{GL}(n, \mathbb{R}) \to \mathbb{R}^\times$ is a homomorphism of multiplicative groups.
2. The exponential map $e: (\mathbb{R}, +) \to (\mathbb{R}^\times, \cdot)$ is a homomorphism.
3. The logarithm is a homomorphism from the multiplicative group of the positive reals $\mathbb{R}_+^\times$ into $(\mathbb{R}, +)$.
4. If H is a subgroup of an abelian group G, then the map $\beta: G \to G/H$ that assigns to each x in G its coset xH in G/H is a homomorphism called the *canonical homomorphism.*
5. If $x \in \mathbb{Z}$, then $m_x: \mathbb{Z} \to \mathbb{Z}$ given by $m_x(y) = xy$ is a homomorphism of additive groups.

The following are all trivial consequences of the definition. If $f: G_1 \to G_2$ is a homomorphism, then

1. $f(e_1) = e_2$ (e_j is the identity of G_j);
2. $f(x^{-1}) = [f(x)]^{-1}$;
3. $f(G_1) = \{f(x) \mid x \in G_1\}$ is a subgroup of G_2, called the *image* of G_1;
4. $\text{Ker}(f) = \{x \in G_1 \mid f(x) = e_2\}$ is a subgroup of G_1, called the *kernel* of f;
5. f is one-to-one (i.e., injective) if and only if $\text{Ker}(f) = \{e_1\}$.

DEFINITION 2.29. A homomorphism $f: G_1 \to G_2$ is called a *monomorphism* if f is one-to-one. Two groups G_1 and G_2 are called *isomorphic*, written as $G_1 \cong G_2$, if there is a monomorphism f from G_1 onto G_2.

EXERCISE 2.30. $G_1 \cong G_2$ is an equivalence relation. [*Hint.* Show that (1) if f is an onto monomorphism then so is f^{-1}; (2) composition of homomorphisms is a homomorphism.]

EXAMPLES 2.31.

1. $(\mathbb{R}_+^\times, \cdot) \cong (\mathbb{R}, +)$.
2. Any group of order 2 is isomorphic to $\mathbb{Z}/2\mathbb{Z}$.
3. Any group of order 4 is isomorphic to $\mathbb{Z}/4\mathbb{Z}$ or $\mathbb{Z}/2\mathbb{Z} \times \mathbb{Z}/2\mathbb{Z}$.
4. Any group of prime order p is isomorphic to $\mathbb{Z}/p\mathbb{Z}$ and hence cyclic.
5. Any finite cyclic group is isomorphic to $\mathbb{Z}/m\mathbb{Z}$ for some m; in particular, $\mu_m \cong \mathbb{Z}/m\mathbb{Z} \cong R_m$.
6. Any infinite cyclic group is isomorphic to $\mathbb{Z}$.

THEOREM 2.32. *Any finitely generated abelian group G is isomorphic to*

$$\underbrace{\mathbb{Z} \times \cdots \times \mathbb{Z}}_{r \text{ copies}} \times \mathbb{Z}/p_1^{n_1}\mathbb{Z} \times \cdots \times \mathbb{Z}/p_k^{n_k}\mathbb{Z},$$

where the p_i are primes, not necessarily distinct.

The non-negative integer r above is called the *rank* of G.

PROOF. See Theorem 17, p. 91 of Ref. 1, or almost any book on group theory or abstract algebra. □

COROLLARY 2.33. *If G is a finite abelian group, then*

$$G \cong \mathbb{Z}/p_1^{n_1}\mathbb{Z} \times \cdots \times \mathbb{Z}/p_k^{n_k}\mathbb{Z}.$$

THEOREM 2.34. (*Isomorphism Theorem*). *If $f: G \to G'$ is a homomorphism of abelian groups, then $G/\mathrm{Ker}(f) \cong f(G)$.*

PROOF (*Sketch*). Let $H = \mathrm{Ker}(f)$ and for a coset xH, put $\bar{f}(xH) = f(x)$. Show that $\bar{f}(x_1H) = \bar{f}(x_2H) \Leftrightarrow x_1H = x_2H$, i.e., $\bar{f}$ does not depend on the coset representative x of the coset xH. Thus $\bar{f}: G/H \to f(G)$. Show that $\bar{f}$ is an onto monomorphism. □

THEOREM 2.35. *Suppose H is a subgroup of an abelian group G and $f: G \to G'$ is a homomorphism of groups. If the index $[G: H]$ is finite, then so are the indices $[f(G): f(H)]$ and $[\mathrm{Ker}(f): \mathrm{Ker}(f) \cap H]$. Moreover,*

$$[f(G): f(H)] = \frac{[G: H]}{[\mathrm{Ker}(f): \mathrm{Ker}(f) \cap H]}.$$

PROOF. Note that if $f_1: G_1 \to G_2$ is an onto homomorphism of groups and G_1 is finite, then from $G_1/\mathrm{Ker}(f_1) \cong G_2$, it follows that

$$|G_2| = \frac{|G_1|}{|\mathrm{Ker}(f_1)|}. \tag{2.4}$$

The map $\hat{f}: G/H \to f(G)/f(H)$ given by $\hat{f}(xH) = f(x)f(H)$ is an onto group homomorphism and by (2.4)

$$[f(G): f(H)] = \frac{[G: H]}{\mathrm{Ker}(\hat{f})}. \tag{2.5}$$

For $x \in \mathrm{Ker}(f)$, $\hat{f}(xH) = f(x)f(H) = f(H)$, so we have a map $\bar{f}: \mathrm{Ker}(f) \to \mathrm{Ker}(\hat{f})$ given by $\bar{f}(x) = xH$. Clearly, $\mathrm{Ker}(\bar{f}) = \mathrm{Ker}(f) \cap H$. If $yH \in \ker(\hat{f})$, then $f(y) = f(h)$ for some $h \in H$, so $x = yh^{-1} \in \mathrm{Ker}(f)$ and $\bar{f}(x) = yh^{-1}H = yH$. This proves that $\bar{f}$ is onto. Therefore, $\mathrm{Ker}(f)/\mathrm{Ker}(f) \cap H \cong \mathrm{Ker}(\hat{f})$ and

$$|\mathrm{Ker}(\hat{f})| = [\mathrm{Ker}(f): \mathrm{Ker}(f) \cap H]. \tag{2.6}$$

Substitute for $|\mathrm{Ker}(\hat{f})|$ from (2.6) in (2.5). □

2.8. Rings

DEFINITION 2.36. A *ring* is a nonempty set A together with two binary operations, written additively and multiplicatively, such that

1. $(A, +)$ is an abelian group;
2. For all x, y, z in A,
   i. $x(yz) = (xy)z$,
   ii. $x(y + z) = xy + xz$ and $(x + y)z = xz + yz$.

DEFINITION 2.37. A ring A is called *commutative* if for all x, y in A, $xy = yx$.

If A has an element 1_A or simply 1, called the *identity* of A, such that for all x in A, $x1 = 1x = x$, we say that A is a *ring with identity* 1_A.

EXAMPLES 2.38.
1. $\mathbb{Z}$ is a commutative ring with identity.
2. $m\mathbb{Z}$ is a commutative ring without identity, if $m > 1$.
3. $M(n, \mathbb{Z})$ is a noncommutative ring with identity, if $n > 1$.
4. $\mathbb{Z}[x_1, \ldots, x_n]$ and $\mathbb{Q}[x_1, \ldots, x_n]$ are commutative rings with identity.
5. Let m be a positive integer and

$$R_m = \mathbb{Z}/m\mathbb{Z} = \{0, 1, 2, \ldots, m - 1\}.$$

We have seen that $(\mathbb{Z}/m\mathbb{Z}, +)$ is an abelian group. Define the product of two remainders r, s to be the remainder of the usual product rs on division by m. Then $\mathbb{Z}/m\mathbb{Z}$ is a commutative ring with identity 1.

EXERCISE 2.39. If x, y are elements of a ring A, show that

1. $0x = 0$;
2. $(-x)y = -xy$;
3. $(-x)(-y) = xy$.

DEFINITION 2.40. A non-empty subset B of a ring A is called a *subring* of A if for all x, y in B,

1. $x - y$ is in B;
2. xy is in B.

EXAMPLES 2.41.
1. $m\mathbb{Z}$ is a subring of $\mathbb{Z}$.
2. $\mathbb{Z}[x_1, \ldots, x_n]$ is a subring of $\mathbb{Q}[x_1, \ldots, x_n]$.

2.9. Ring Homomorphisms

DEFINITION 2.42. Suppose A, B are two rings. A map $f: A \to B$ is called a (*ring*) *homomorphism* if for all x, y in A,

1. $f(x + y) = f(x) + f(y)$;
2. $f(xy) = f(x)f(y)$.

If $A = B$, f is called an *endomorphism* of A. Two rings A and B are *isomorphic*, written as $A \cong B$, if there is a bijective ring homomorphism $f: A \to B$.

EXAMPLE 2.43. Let $A = \mathbb{Z}$, $B = \mathbb{Z}/m\mathbb{Z}$. Denote by $\bar{x}$, or $r_m(x)$, the remainder of x on division by m. Then $r_m: \mathbb{Z} \to \mathbb{Z}/m\mathbb{Z}$ is a ring homomorphism called the *reduction modulo m*. The proof follows at once from Theorem 1.27.

2.10. Fields

DEFINITION 2.44. A *field* is a commutative ring K with identity $1 \neq 0$, such that each nonzero element x in K has a multiplicative inverse x^{-1}.

It can be checked that x^{-1} is unique.

EXAMPLES 2.45.
1. $K = \mathbb{Q}, \mathbb{R}, \mathbb{C}$ are all fields.
2. $\mathbb{Z}$ is not a field, because no nonzero x (except $x = \pm 1$) has a multiplicative inverse.

3. Let

$$K = \left\{ \frac{f(\mathbf{x})}{g(\mathbf{x})} \,\middle|\, f(\mathbf{x}), g(\mathbf{x}) \in \mathbb{Q}[x_1, \ldots, x_n] \text{ and } g(\mathbf{x}) \neq 0 \right\}.$$

Then K is a field under the usual operation of addition and multiplication. The elements of K are called *rational functions.* Note that

$$K = \left\{ \frac{f(\mathbf{x})}{g(\mathbf{x})} \,\middle|\, f(\mathbf{x}), g(\mathbf{x}) \in \mathbb{Z}[x_1, \ldots, x_n], g(\mathbf{x}) \neq 0 \right\}.$$

2.11. Finite Fields

Let $p \geq 2$ be a prime and let $\mathbb{F}_p$ denote the ring $\mathbb{Z}/p\mathbb{Z}$. To show that $\mathbb{F}_p$ is a field, we must show that for each $x \neq 0$, x^{-1} exists. This follows at once from Theorem 1.30. We also give here another proof of this fact.

For $0 < x < p$, no two of $0, x, 2x, \ldots, (p-1)x$ leave the same remainder on division by p, because otherwise, $ix \equiv jx \pmod{p}$ with $0 \leq j < i < p$ implies that $p \mid (i-j)x$. But $(p, x) = 1$, so $p \mid i-j$, which is impossible. Thus $xy \equiv 1 \pmod{p}$ for some y $(1 < y < p)$ and $y = x^{-1}$.

Note that $\mathbb{F}_p$ is a *finite field* of p elements. For any field K, the set $K^\times$ of its nonzero elements is a multiplicative group. In particular, $\mathbb{F}_p^\times$ is a group of $p-1$ elements. This group $\mathbb{F}_p^\times$ has an important subgroup, viz.,

$$\mathbb{F}_p^{\times 2} = \{x^2 \mid x \in \mathbb{F}_p^\times\}.$$

We shall show that if $p > 2$, the index $[\mathbb{F}_p^\times : \mathbb{F}_p^{\times 2}]$ is always 2.

2.12. Polynomials over Rings

Let A be a commutative ring with identity 1. The set $A[x]$ consisting of *polynomials*

$$f(x) = a_0 + a_1 x + \cdots + a_n x^n \qquad (a_j \in A)$$

over A is a ring under the usual definition of addition and multiplication of polynomials. If $a_n \neq 0$, the *degree* $\deg f(x)$ of $f(x)$ is defined to be n. Moreover, if A is a field and $d(x) \neq 0$ is another polynomial over A, we can use the "synthetic division" or the *division algorithm* to write

$$f(x) = q(x)d(x) + r(x), \qquad \deg r(x) < \deg d(x);$$

in one and only one way, with *quotient* $q(x)$ and the *remainder* $r(x)$ as polynomials over A. [We follow the convention that $\deg(0) = -\infty$.]

EXERCISE 2.46. Perform the synthetic division with the polynomials

$$f(x) = 3x^5 + 4x^4 + x^3 + 3x + 1$$

and

$$d(x) = 2x + 3$$

over the field $\mathbb{F}_5$ to find the quotient $q(x)$ and the remainder $r(x)$ in $\mathbb{F}_5[x]$.

DEFINITION 2.47. Suppose K is a field, $f(x) \in K[x]$ and $\alpha \in K$. We say that α is a *root* of $f(x)$ if $f(\alpha) = 0$.

If α is a root of $f(x)$, it is clear from the division algorithm with $d(x) = x - \alpha$ that

$$f(x) = (x - \alpha)q(x),$$

i.e., $x - \alpha$ *divides* $f(x)$. Also $\deg q(x) = \deg f(x) - 1$. Applying the same argument to $q(x)$ we obtain the following theorem.

THEOREM 2.48. *A polynomial over a field K of degree n cannot have more than n roots in K.*

COROLLARY 2.49. $\mathbb{F}_p^\times$ *is a cyclic group of order* $p - 1$.

PROOF. If $\mathbb{F}_p^\times$ is not cyclic, then $\operatorname{ord}(x) < p - 1$ for each x in $\mathbb{F}_p^\times$ [cf. Exercise 2.26(2)]. Let

$$r = \max_{x \in \mathbb{F}_p^\times} \operatorname{ord}(x) < p - 1.$$

First we show that $\operatorname{ord}(x) \mid r$, for each x in $\mathbb{F}_p^\times$. If not, let $\mathbb{F}_p^\times$ have an element of order s, $1 < s < r$, such that r is not divisible by s. By Exercise 2.22(1), $\mathbb{F}_p^\times$ has an element of order $[r, s]$ which is clearly larger than r. This contradicts the maximality of r.

Now it is obvious that each element of $\mathbb{F}_p^\times$ is a root of the polynomial

$$f(x) = x^r - 1$$

showing that $f(x)$ has more roots in $\mathbb{F}_p^\times$ than its degree. This contradiction completes the proof. □

THEOREM 2.50. $\mathbb{F}_p^{\times 2}$ *is a subgroup of* $\mathbb{F}_p^\times$ of index

$$[\mathbb{F}_p^\times : \mathbb{F}_p^{\times 2}] = \begin{cases} 1, & \text{if } p = 2, \\ 2, & \text{if } p > 2. \end{cases}$$

PROOF. For $p = 2$, there is nothing to prove. So let $p > 2$. First we note that in $\mathbb{F}_p^\times$, 1 and -1 are the only roots of $x^2 - 1$ and form a group of order 2.

The map $\psi: \mathbb{F}_p^\times \to \mathbb{F}_p^{\times 2}$ given by $\psi(x) = x^2$ is an onto group homomorphism with $\operatorname{Ker}(\psi) = \{\pm 1\}$. Hence by Theorem 2.34, $\mathbb{F}_p^{\times 2} \cong \mathbb{F}_p^\times/\{\pm 1\}$, and this proves the theorem. □

EXAMPLES 2.51.
1. $\mathbb{F}_5^{\times 2} = \{1, 4\}$.
2. $\mathbb{F}_{11}^{\times 2} = \{1, 3, 4, 5, 9\}$.

REMARK 2.52. Suppose p is a prime and $(\alpha_1, \ldots, \alpha_n)$ is an integer solution of

$$f(x_1, \ldots, x_n) = 0, \tag{2.7}$$

where

$$f(x_1, \ldots, x_n) \in \mathbb{Z}[x_1, \ldots, x_n].$$

Suppose $\bar{f}(x_1, \ldots, x_n) \in \mathbb{F}_p[x_1, \ldots, x_n]$ is obtained on replacing the coefficients of $f(x_1, \ldots, x_n)$ by their remainders on division by p. Then by Corollary 1.28, $(\bar{\alpha}_1, \ldots, \bar{\alpha}_n)$ is a solution in $\mathbb{F}_p$ of the equation

$$\bar{f}(x_1, \ldots, x_n) = 0,$$

called the *reduction of* (2.7) *mod p.*

This way we can show, for example, that the equation

$$11x^2 - 10y^2 = 12$$

has no solution in integers. Because if it did, then its reduction mod $p = 5$, i.e.,

$$x^2 = 2$$

must have a solution too in $\mathbb{F}_5$, i.e., 2 must be a square in $\mathbb{F}_5^\times$, which is not the case (Example 2.51).

Reference

1. H. Zassenhaus, *The Theory of Groups*, Chelsea, New York (1949).

3

Representation of Integers by Forms

3.1. Introduction

The squares (of integers), namely

$$0, 1, 4, 9, 16, 25, \ldots$$

are very sparse. However, the integers that are sums of two squares occur more frequently:

$$\begin{aligned} 1 &= 1^2 + 0^2, \\ 2 &= 1^2 + 1^2, \\ 4 &= 2^2 + 0^2, \\ 5 &= 2^2 + 1^2, \\ 8 &= 2^2 + 2^2, \\ 9 &= 3^2 + 0^2, \\ 10 &= 3^2 + 1^2, \\ &\vdots \end{aligned}$$

But still there are integers such as $3, 6, 7, \ldots$ that cannot be written as a sum of two squares. Thus one may ask the following question:

How do we decide whether a positive integer n is a sum of two squares or not; i.e., for which $n > 0$, *does the equation*

$$n = x_1^2 + x_2^2$$

have a solution in integers?

Sums of three squares fill some more gaps, but still we do not get all the positive integers. One may further ask:

Is there a positive integer g, such that the sums of g squares account for all the positive integers? Further, if there is such an integer g, what is the least value of g that suffices?

This is a special case of Waring's problem:

WARING'S PROBLEM (1770). Suppose $k > 1$ is an integer. Does there exist a positive integer g, such that every positive integer n is a sum of g kth powers, i.e., such that the equation

$$n = x_1^k + \cdots + x_g^k \tag{3.1}$$

has a solution in integers for all $n > 0$?

This question was answered by Hilbert in the affirmative (cf. Ref. 3). For a simpler proof see Ref. 1. The next obvious question is given $k > 1$, what is the least value $g(k)$ of g that suffices for (3.1) to have a solution for all $n > 0$? Since 7 is not a sum of three squares, $g(2) > 3$. It was Lagrange who showed that $g(2) = 4$, i.e., every positive integer is a sum of four squares.

Waring's problem is a special case of a more general problem: to study the representation of numbers by forms.

DEFINITION 3.1. Let d be a positive integer. A *form of degree d* over a ring A is a homogeneous polynomial

$$f(\mathbf{x}) = f(x_1, \ldots, x_n)$$

with coefficients in A and of degree d.

This means that for the parameter t,

$$\begin{aligned} f(t\mathbf{x}) &= f(tx_1, \ldots, tx_n) \\ &= t^d f(\mathbf{x}). \end{aligned} \tag{3.2}$$

If for non-negative integers $\alpha_1, \ldots, \alpha_n$ we define the degree of the *monomial* $x_1^{\alpha_1} \cdots x_n^{\alpha_n}$ by

$$\deg(x_1^{\alpha_1} \cdots x_n^{\alpha_n}) = \alpha_1 + \cdots + \alpha_n,$$

then (3.2) is equivalent to saying that each term of $f(\mathbf{x})$ is of degree d.

For a vector

$$\boldsymbol{\alpha} = (\alpha_1, \ldots, \alpha_n)$$

with integer coordinates $\alpha_j \geq 0$, $j = 1, \ldots, n$, we put

$$|\boldsymbol{\alpha}| = \alpha_1 + \cdots + \alpha_n$$

and

$$\mathbf{x}^{\boldsymbol{\alpha}} = x_1^{\alpha_1} \cdots x_n^{\alpha_n}.$$

A form of degree d over A then may be conveniently written as

$$f(\mathbf{x}) = \sum_{|\boldsymbol{\alpha}|=d} c_{\boldsymbol{\alpha}} x^{\boldsymbol{\alpha}}, \tag{3.3}$$

the coefficients $c_{\boldsymbol{\alpha}}$ being in A.

DEFINITION 3.2. A form of degree d is called a *linear* or a *quadratic form* according as $d = 1$ or 2. If $d > 2$, $f(\mathbf{x})$ is called a *higher form* or a *form of higher degree.*

DEFINITION 3.3. Suppose $f(\mathbf{x})$ is a form of degree d with coefficients in $\mathbb{Z}$. We say that $f(\mathbf{x})$ *represents* an integer n if there is a vector

$$\mathbf{a} = (a_1, \ldots, a_n)$$

with integer coordinates, such that $n = f(\mathbf{a})$. If $n = 0$, $\mathbf{a}$ is called a *zero* of $f(\mathbf{x})$.

As has already been mentioned, and will be proved in this chapter, every integer $n \geq 0$ is represented by the quadratic form

$$f(\mathbf{x}) = x_1^2 + x_2^2 + x_3^2 + x_4^2.$$

On the other hand, the form

$$h(\mathbf{x}) = x_1^2 + x_2^2 + x_3^2$$

in three variables does not represent 7.

We will deal only with quadratic forms. The theory of higher forms is still at a developing stage (cf. Refs. 4 and 5). First we establish the so-called law of *quadratic reciprocity.* It was first conjectured by Euler in 1783 and in the form we shall present here by Legendre in 1785. Gauss rediscovered it at the age of eighteen and gave its first proof a year later in 1796. Later on he gave six more entirely different proofs. Now there are well over 50 proofs of the law of quadratic reciprocity. However, most of them are based on more or less the seven proofs given by Gauss.

3.2. Quadratic Reciprocity

DEFINITION 3.4. Let $m > 1$ be a fixed integer. An integer a with $(a, m) = 1$ is called a *quadratic residue modulo m* if the congruence equation

$$x^2 \equiv a (\text{mod } m) \tag{3.4}$$

has a solution. If (3.4) has no solution, a is called a *quadratic nonresidue modulo m.*

The term "quadratic nonresidue" is traditional, even though it seems to be more correct to use "nonquadratic residue." Note that if $a \equiv b \pmod{m}$, then a is a quadratic residue modulo m if and only if b is a quadratic residue modulo m. For $m = p$, p an odd prime, a is a quadratic residue mod p if and only if its remainder r $(0 < r < p)$ on division by p is in $\mathbb{F}_p^{\times 2}$.

For the rest of this section we fix an odd prime p. Then $(p-1)/2$ is an integer. The function

$$\sigma: \mathbb{F}_p^\times \to \mathbb{F}_p^\times$$

given by

$$y = \sigma(x) = x^{(p-1)/2}$$

is a group homomorphism. For each x in $\mathbb{F}_p$, $\sigma(x) = \pm 1$, because it is a root of the polynomial $y^2 = 1$. Moreover,

1. $\sigma(x) = -1$ for some x in $\mathbb{F}_p^\times$, because otherwise $x^{(p-1)/2} - 1$ will have more roots than its degree;
2. If $x = t^2 \in \mathbb{F}_p^{\times 2}$, then

$$\begin{aligned}\sigma(x) &= x^{(p-1)/2} = (t^2)^{(p-1)/2} \\ &= t^{p-1} = 1.\end{aligned}$$

Putting 1 and 2 together, we get

$$\mathbb{F}_p^{\times 2} \subseteq \ker(\sigma) \subsetneqq \mathbb{F}_p^\times.$$

The index

$$[\mathbb{F}_p^\times : \ker(\sigma)] \geq 2. \tag{3.5}$$

Since

$$2 = [\mathbb{F}_p^\times : \mathbb{F}_p^{\times 2}] = [\mathbb{F}_p^\times : \ker(\sigma)][\ker(\sigma) : \mathbb{F}_p^{\times 2}],$$

equality must hold in (3.5) and we must have

$$[\ker(\sigma) : \mathbb{F}_p^{\times 2}] = 1.$$

Therefore,

$$\ker(\sigma) = \mathbb{F}_p^{\times 2}.$$

DEFINITION 3.5. The *Legendre symbol* is the group homomorphism

$$\left(\frac{-}{p}\right) = \sigma: \mathbb{F}_p^\times \to \{\pm 1\}$$

of the multiplicative groups.

This means that if $a \in \mathbb{F}_p^\times$,

$$\left(\frac{a}{p}\right) \stackrel{\text{def}}{=} \sigma(a) = a^{(p-1)/2}$$

$$= \begin{cases} 1, & \text{if } a \text{ is a quadratic residue mod } p, \\ -1, & \text{otherwise.} \end{cases}$$

The Legendre symbol may also be defined on the set of those integers that are coprime to p as a composition of σ and the reduction mod p, i.e., applying σ to the remainder of a on division by p.

In particular,

$$\left(\frac{-1}{p}\right) = (-1)^{(p-1)/2}. \tag{3.6}$$

In view of σ being a group homomorphism, we have

$$\left(\frac{ab}{p}\right) = \left(\frac{a}{p}\right)\left(\frac{b}{p}\right). \tag{3.7}$$

So far we have kept the odd prime p fixed. Now suppose that $q \neq p$ is another odd prime. Then q (more precisely its reduction mod p) is in $\mathbb{F}_p^\times$. Similarly $p \in \mathbb{F}_q^\times$. How is q being a square in $\mathbb{F}_p^\times$ related to p being a square in $\mathbb{F}_p^\times$? The answer is the law of quadratic reciprocity (cf. Theorem 3.10). Among the seven gaussian proofs the third and the fifth are the most elementary and simple. Both are based on Gauss's lemma. We shall give his third proof. For more proofs see Ref. 8.

THEOREM 3.6. (*Gauss's lemma*). *Write* $\mathbb{F}_p^\times = X \cup Y$ *as a disjoint union of the sets*

$$X = \{1, 2, \ldots, (p-1)/2\}$$

and

$$Y = \{(p+1)/2, \ldots, p-1\}.$$

For $a \in \mathbb{F}_p^\times$ *let* $aX = \{ax \mid x \in X\}$ *and denote by* g *the number of elements in* $aX \cap Y$. *Then*

$$\left(\frac{a}{p}\right) = (-1)^g. \tag{3.8}$$

PROOF. First note that the function

$$m_a : \mathbb{F}_p^\times \to \mathbb{F}_p^\times$$

given by $m_a(x) = ax$ permutes the elements of $\mathbb{F}_p^\times$ (Theorem 2.10). So if

$$aX \cap X = \{x_1, \ldots, x_k\}$$

and

$$aX \cap Y = \{y_1, \ldots, y_g\}$$

then

$$g + k = (p-1)/2. \tag{3.9}$$

If

$$Z = \{x_1, \ldots, x_k, p - y_1, \ldots, p - y_g\},$$

then $Z \subset X$. Moreover, $x_1, \ldots, x_k$, $p - y_1, \ldots, p - y_g$ are all distinct elements of $\mathbb{F}_p$. To prove this all we have to show is that $x_i \neq p - y_j$ for all i, j. Suppose the contrary, i.e., $x_i = p - y_j$ for some i, j. Noting that p is zero in $\mathbb{F}_p$ (and performing the operations in $\mathbb{F}_p$) we get $x_i + y_j = 0$. But $x_i = ar$ and $y_j = as$ for some r, s $(1 \leq r, s \leq (p-1)/2)$. Hence $a(r+s) = 0$. Since $a \neq 0$, we must have $r + s = 0$, i.e., $p \mid r + s$. But this is impossible because $2 \leq r + s \leq p - 1$. This and (3.9) now imply that the two sets X and Z are equal and hence (in $\mathbb{F}_p$)

$$\begin{aligned}
1 \cdot 2 \cdots \frac{p-1}{2} &= x_1 \ldots x_k (p - y_1) \ldots (p - y_g) \\
&= (-1)^g x_1 \ldots x_k y_1 \ldots y_g \\
&= (-1)^g a \cdot 2a \cdot 3a \ldots \frac{p-1}{2} a \\
&= (-1)^g a^{(p-1)/2} \cdot 1 \cdot 2 \cdot 3 \cdots \frac{p-1}{2}.
\end{aligned}$$

This gives

$$(-1)^g a^{(p-1)/2} = 1$$

or

$$(-1)^g = a^{(p-1)/2} = \left(\frac{a}{p}\right). \qquad \square$$

NOTATION 3.7. For a real number α, we shall write $[\alpha]$ for the largest integer not exceeding α. As examples $[13/3] = 4$, $[-4/3] = -2$.

COROLLARY 3.8. *For an odd prime p,*

$$\left(\frac{2}{p}\right) = (-1)^{(p^2-1)/8}.$$

PROOF. First note that $(p^2-1)/8$ is an integer. In fact it is even or odd according as $p \equiv 1, 7 \pmod 8$ or $p \equiv 3, 5 \pmod 8$. To see this write

$$p = 8m + r \qquad (r = 1, 3, 5, 7). \tag{3.10}$$

Then

$$\begin{aligned}\frac{p^2-1}{8} &= \frac{(8m+r)^2-1}{8}\\ &= 2n + \frac{r^2-1}{8}\end{aligned}$$

for some n, which is obviously even for $r = 1, 7$ and odd for $r = 3, 5$.

For $1 \le x \le (p-1)/2$, $2x$ never exceeds $p-1$. Thus g in Theorem 3.6 is the number of $2x$, $1 \le x \le (p-1)/2$, that exceed $(p-1)/2$, i.e., the number of x, $1 \le x \le (p-1)/2$, such that $x > (p-1)/4$. Therefore

$$g = \frac{p-1}{2} - \left[\frac{p-1}{4}\right].$$

Using (3.10), this gives

$$\begin{aligned}g &= 4m + \frac{r-1}{2} - \left[2m + \frac{r-1}{4}\right]\\ &= 4m + \frac{r-1}{2} - 2m - \left[\frac{r-1}{4}\right]\\ &= 2m + \frac{r-1}{2} - \left[\frac{r-1}{4}\right],\end{aligned}$$

which is even if $r = 1, 7$ and odd if $r = 3, 5$. Thus g and $(p^2-1)/8$ have the same parity and

$$\left(\frac{2}{p}\right) = (-1)^g = (-1)^{(p^2-1)/8}. \qquad \square$$

For the proof of the law of quadratic reciprocity, we need the following combinatorial result.

LEMMA 3.9. *If p and q are distinct odd primes, then*

$$\sum_{j=1}^{(p-1)/2}\left[\frac{jq}{p}\right] + \sum_{j=1}^{(q-1)/2}\left[\frac{jp}{q}\right] = \frac{p-1}{2}\cdot\frac{q-1}{2}.$$

Proof. If we put

$$s(p, q) = \sum_{j=1}^{(p-1)/2} \left[\frac{jq}{p}\right],$$

we must show that

$$s(p, q) + s(q, p) = \frac{(p-1)(q-1)}{4}. \tag{3.11}$$

It is easy to see that for each $j = 1, \ldots, (p-1)/2$, $[jq/p]$ is the number of integers in the open interval

$$(0, jq/p) = \{x \in \mathbb{R} \mid 0 < x < jq/p\}.$$

Hence for each j, $[jq/p]$ is the number of those lattice points (that is, the points with integer coordinates) on the line $x = j$ that lie below the line

$$y = \frac{q}{p}x \tag{3.12}$$

and above (but not on) the line $y = 0$. Note that there are no lattice points on (3.12) with $0 < x \le (p-1)/2$. Thus $s(p, q)$, being the sum of $[jq/p]$ for $j = 1, 2, \ldots, (p-1)/2$, is the number of those lattice points in the interior of (that is, inside but not on) the rectangle OACB that are below the line (3.12) (see Fig. 3.1).

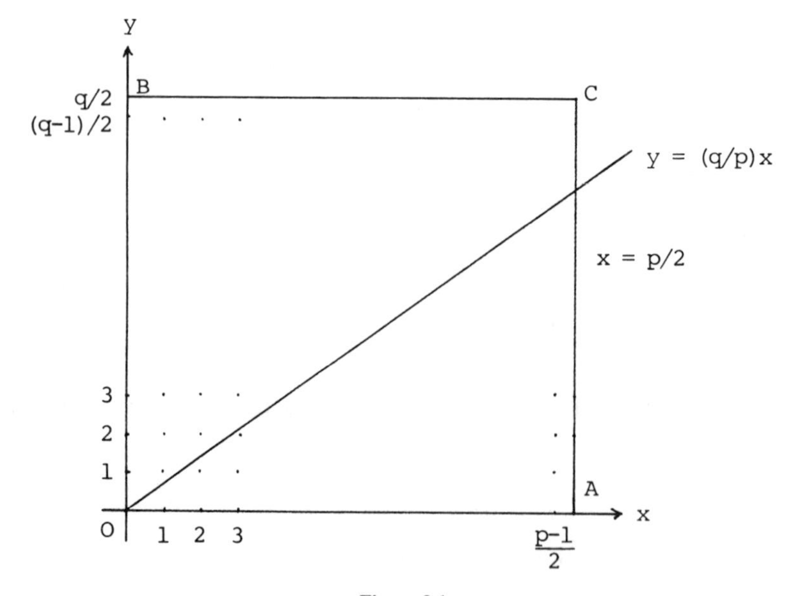

Figure 3.1.

Similarly $s(q, p)$ is the number of those lattice points in the interior of OACB that are above (3.12). Thus $s(p, q) + s(q, p)$ is the total number of the lattice points in the interior of OACB, which is clearly

$$\frac{p-1}{2} \cdot \frac{q-1}{2}. \qquad \square$$

THEOREM 3.10 (*Law of Quadratic Reciprocity*). *If p and q are distinct odd primes, then*

$$\left(\frac{p}{q}\right)\left(\frac{q}{p}\right) = (-1)^{(p-1)/2 \cdot (q-1)/2}.$$

PROOF. Let the notations be as in the proof of Theorem 3.6, but regard x_i, y_j as integers (not elements of $\mathbb{F}_p$). Put

$$\alpha = \sum_{j=1}^{k} x_j, \qquad \beta = \sum_{j=1}^{g} y_j.$$

Using the formula

$$1 + 2 + \cdots + N = \frac{N(N+1)}{2},$$

we get

$$\sum_{x \in X} x = 1 + 2 + \cdots + \frac{p-1}{2}$$

$$= \frac{p^2 - 1}{8}.$$

And as in the proof of Theorem 3.6 with $a = q$,

$$\sum_{z \in Z} z = \sum_{j=1}^{k} x_j + \sum_{j=1}^{g} p - y_j$$

$$= \alpha - \beta + pg.$$

But $Z = X$, so

$$\frac{p^2 - 1}{8} = \alpha - \beta + pg. \tag{3.13}$$

Now for $j = 1, \ldots, (p-1)/2$, let t_j denote the remainder of jq on division by p. Clearly the quotient is $[jq/p]$ and

$$jq = [jq/p]p + t_j. \tag{3.14}$$

Taking the sum of (3.14) for $j = 1, \ldots, (p-1)/2$, we get

$$\frac{q(p^2-1)}{8} = ps(p,q) + \sum_{j=1}^{(p-1)/2} t_j$$

$$= ps(p,q) + \sum_{j=1}^{k} x_j + \sum_{j=1}^{g} y_j$$

or

$$\frac{q(p^2-1)}{8} = ps(p,q) + \alpha + \beta.$$

Substituting for α from (3.13) in the above equation, we get

$$\frac{(q-1)(p^2-1)}{8} = p(s(p,q) - g) + 2\beta. \tag{3.15}$$

Since p, q are odd primes and $(p^2-1)/8$ is an integer, it is clear from (3.15) that

$$s(p,q) - g \equiv 0 \pmod{2}.$$

Hence

$$\left(\frac{q}{p}\right) = (-1)^g = (-1)^{s(p,q)}. \tag{3.16}$$

Interchanging the role of p and q,

$$\left(\frac{p}{q}\right) = (-1)^{s(q,p)}. \tag{3.17}$$

Multiplying (3.16) and (3.17) and using Lemma 3.9, we get

$$\left(\frac{p}{q}\right)\left(\frac{q}{p}\right) = (-1)^{s(p,q)+s(q,p)}$$

$$= (-1)^{[(p-1)/2][(q-1)/2]}. \qquad \square$$

EXAMPLE. Let $p = 1009$. To decide whether 45 is square in $\mathbb{F}_p^\times$ or not, we note that $45 = 3^2 \cdot 5$. Therefore

$$\left(\frac{45}{1009}\right) = \left(\frac{3^2}{1009}\right)\left(\frac{5}{1009}\right) = \left(\frac{5}{1009}\right)$$

$$= \left(\frac{1009}{5}\right)(-1)^{(1009-1)/2 \cdot (5-1)/2}$$

$$= \left(\frac{1009}{5}\right) = \left(\frac{9}{5}\right) = 1.$$

EXERCISES.

1. Compute $(-\frac{30}{257})$ and $(\frac{1987}{1997})$.
2. Find all the primes p for which $(-10/p) = 1$.
3. Find all the primes p for which $(5/p) = -1$.

3.3. Some Special Quadratic Forms

Lagrange proved that every non-negative integer is a sum of four squares, i.e., the quadratic form

$$f(\mathbf{x}) = x_1^2 + x_2^2 + x_3^2 + x_4^2 \tag{3.18}$$

represents every non-negative integer. Euler discovered the identity

$$\begin{aligned}(x_1^2 + x_2^2 + x_3^2 + x_4^2)&(y_1^2 + y_2^2 + y_3^2 + y_4^2)\\ &= (x_1y_1 + x_2y_2 + x_3y_3 + x_4y_4)^2 + (x_1y_2 - x_2y_1 + x_3y_4 - x_4y_1)^2\\ &\quad + (x_1y_3 - x_3y_1 + x_4y_2 - x_2y_4)^2 + (x_1y_4 - x_4y_1 + x_2y_3 - x_3y_2)^2,\end{aligned} \tag{3.19}$$

which shows that if n_j is a sum of two squares, $j = 1, 2$, then so is n_1n_2. This reduces the task of proving the above statement to the proof of the fact that every prime p is a sum of four squares. Clearly

$$2 = 1^2 + 1^2 + 0^2 + 0^2, \tag{3.20}$$

so we may assume that p is odd.

THEOREM 3.11 (*Lagrange*). *Every positive integer is a sum of four squares.*

PROOF (*Euler*). In view of (3.19) and (3.20) it suffices to show that every odd prime p is a sum of four squares. We shall prove it in two steps:

1. There are integers m and x_j $(1 \le j \le 4)$ with

$$mp = x_1^2 + x_2^2 + x_3^2 + x_4^2 \qquad (1 \le m < p). \tag{3.21}$$

2. If m is the smallest integer satisfying (3.21), then $m = 1$.

To prove 1, consider the sets

$$X = \{x^2 | x = 0, 1, \ldots, (p-1)/2\}$$

and

$$Y = \{-x^2 - 1 | x = 0, 1, \ldots, (p-1)/2\}.$$

No two elements of X are congruent modulo p. If they were, say

$$x_1^2 \equiv x_2^2 \pmod p, x_1 > x_2,$$

then $p | x_1^2 - x_2^2 = (x_1 + x_2)(x_1 - x_2)$. Therefore either $p | x_1 + x_2$ or $p | x_1 - x_2$. But this is impossible, because $1 \le x_1 \pm x_2 \le p - 1$. Similarly no two elements of Y are congruent modulo p. Since

$$|X| + |Y| > p,$$

where $|X|$ denotes the number of elements in X, we must have

$$x^2 \equiv -y^2 - 1 \pmod p,$$

for some x, y $[1 \le x, y \le (p-1)/2]$, i.e.,

$$mp = x^2 + y^2 + 1$$

for an integer m.

Clearly

$$\begin{aligned} 1 \le m &= (1/p)(x^2 + y^2 + 1) \\ &\le \frac{1}{p}\left[2 \cdot \left(\frac{p-1}{2}\right)^2 + 1\right] \\ &= \frac{p-1}{p} \cdot \frac{p-1}{2} + \frac{1}{p} \\ &< \frac{p-1}{2} + \frac{1}{p} \\ &< p. \end{aligned}$$

To prove 2, first note that if m is even then either none, two, or four of the x_i are even. When exactly two x_i are even, we assume that these are x_1 and x_2. So in any case $x_1 \pm x_2$ and $x_3 \pm x_4$ are all even and we have

$$\left(\frac{x_1 + x_2}{2}\right)^2 + \left(\frac{x_1 - x_2}{2}\right)^2 + \left(\frac{x_3 + x_4}{2}\right)^2 + \left(\frac{x_3 - x_4}{2}\right)^2 = \frac{m}{2}p.$$

But this contradicts the minimality of m. So m must be odd.

If $m = 1$, there is nothing to prove. So let $3 \leq m < p$. Now choose y_i such that

$$x_i \equiv y_i \pmod{m}, \ -\frac{m-1}{2} \leq y_i \leq \frac{m-1}{2}, \tag{3.22}$$

$i = 1, 2, 3, 4$. In view of (3.21) and (3.22), it is clear that

$$y_1^2 + y_2^2 + y_3^2 + y_4^2 \equiv 0 \pmod{m},$$

i.e.,

$$mn = y_1^2 + y_2^2 + y_3^2 + y_4^2 \tag{3.23}$$

for some n. Moreover,

$$0 \leq n \leq \frac{4}{m}\left(\frac{m-1}{2}\right)^2 < m.$$

The integer $n \neq 0$, because otherwise, $y_j = 0$ for all j. This would imply that $x_j = 0 \pmod{m}$, $j = 1, 2, 3, 4$ and hence

$$\begin{aligned} mp &= x_1^2 + x_2^2 + x_3^2 + x_4^2 \\ &= 0 \pmod{m^2}. \end{aligned}$$

This implies that $p \equiv 0 \pmod{m}$, which is impossible, because $3 \leq m < p$. So $n \geq 1$.

Multiplying (3.21) and (3.23) and abbreviating the right-hand side of (3.19), we get

$$\begin{aligned} m^2np &= (x_1^2 + x_2^2 + x_3^2 + x_4^2)(y_1^2 + y_2^2 + y_3^2 + y_4^2) \\ &= z_1^2 + z_2^2 + z_3^2 + z_4^2. \end{aligned}$$

Using (3.22) we see that $m \mid z_j$, $j = 2, 3, 4$ and hence $m \mid z_1$ also. So

$$np = \left(\frac{z_1}{m}\right)^2 + \left(\frac{z_2}{m}\right)^2 + \left(\frac{z_3}{m}\right)^2 + \left(\frac{z_4}{m}\right)^2,$$

contradicting the minimality of m. This proves the theorem.

Exercise 3.12. Verify identity (3.19).

Remark 3.13. The identity (3.19) can be explained in a very nice way. We define the *ring* $\mathbb{H}$ *of real quaternions*, first discovered by Hamilton, as follows:

Fix three symbols i, j, k. Consider the set $\mathbb{H}$ consisting of the formal symbols (called the *quaternions*)

$$x_0 + x_1 i + x_2 j + x_3 k; \ x_r \in \mathbb{R}, \qquad r = 0, 1, 2, 3.$$

Let

$$\alpha = x_0 + x_1 i + x_2 j + x_3 k$$

and $\qquad\qquad\qquad\qquad\qquad\qquad\qquad\qquad\qquad\qquad\qquad\qquad (*)$

$$\beta = y_0 + y_1 i + y_2 j + y_3 k.$$

By definition $\alpha = \beta$ if and only if $x_r = y_r$ for all r.

The set $\mathbb{H}$ becomes an abelian group if we define an addition on $\mathbb{H}$ by

$$\begin{aligned}(x_0 + x_1 i + x_2 j + x_3 k) &+ (y_0 + y_1 i + y_2 j + y_3 k) \\ &= (x_0 + y_0) + (x_1 + y_1)i + (x_2 + y_2)j + (x_3 + y_3)k.\end{aligned}$$

The additive identity is

$$0 = 0 + 0i + 0j + 0k.$$

[In fact, $\mathbb{H}$ is a vector space of dimension 4 over the field $\mathbb{R}$ of real numbers, if the scalar multiplication is defined in an obvious way: for $x \in \mathbb{R}$,

$$x(x_0 + x_1 i + x_2 j + x_3 k) = xx_0 + xx_1 i + xx_2 j + xx_3 k.]$$

We now define a multiplication on $\mathbb{H}$. Let α, β be as in $(*)$. Put

$$\begin{aligned}\alpha\beta = (x_0y_0 - x_1y_1 - x_2y_2 - x_3y_3) &+ (x_0y_1 + x_1y_0 + x_2y_3 - x_3y_2)i \\ + (x_0y_2 + x_2y_0 + x_3y_1 - x_1y_3)j &+ (x_0y_3 + x_3y_0 + x_1y_2 - x_2y_1)k.\end{aligned}$$

This amounts to multiplying quaternions formally, keeping in mind that

$$i^2 = j^2 = k^2 = -1;$$

$$ij = -ji = k, \qquad jk = -kj = i, \qquad ki = -ik = j$$

and that for $x \in \mathbb{R}$,

$$xi = ix, \qquad xj = jx, \qquad xk = kx.$$

These two operations make $\mathbb{H}$ into a noncommunicative ring with identity 1. Moreover, $\mathbb{H}$ contains the field $\mathbb{C}$ of complex numbers as a subring. The notion of (complex) conjugation on $\mathbb{C}$ can be extended to that of conjugation on $\mathbb{H}$. If α is as in $(*)$, we define its *conjugate* $\bar{\alpha}$ by

$$\bar{\alpha} = x_0 - x_1 i - x_2 j - x_3 k.$$

The *norm* $N(\alpha)$ or the *length* $|\alpha|$ of α can now be defined in a natural way:

$$\begin{aligned}N(\alpha) = |\alpha|^2 &= \alpha\bar{\alpha} \\ &= x_0^2 + x_1^2 + x_2^2 + x_3^2.\end{aligned}$$

Note that $N(0) = 0$. If $\alpha \neq 0$, then $N(\alpha)$ is a (strictly) positive real number. Thus it follows that each $\alpha \neq 0$ has a multiplicative inverse α^{-1} given by

$$\alpha^{-1} = \frac{1}{N(\alpha)}\,\bar{\alpha}.$$

The identity (3.19) can be written, with a slight change of notation, as

$$N(\alpha\beta) = N(\alpha)N(\beta). \tag{3.24}$$

Now we study the integers represented by the quadratic form

$$f(\mathbf{x}) = x_1^2 + x_2^2.$$

We begin with the following theorem.

THEOREM 3.14 (*Euler*). *Every prime* $p \equiv 1 \pmod 4$ *is a sum of two squares.*

PROOF. As in the proof of Theorem 3.11, it suffices to show the following:

1. There are integers m, x_1, and x_2 such that

$$mp = x_1^2 + x_2^2, \qquad 1 \leq m < p. \tag{3.25}$$

2. If m is the smallest integer satisfying (3.25), then $m = 1$.

To prove (1), note that since $p \equiv 1 \pmod 4$, the Legendre symbol

$$\left(\frac{-1}{p}\right) = (-1)^{(p-1)/2} = 1.$$

Therefore $-1 \in \mathbb{F}_p^{\times 2}$, i.e., $mp = x^2 + 1$ for some m in $\mathbb{Z}$ and $1 < x < p$. Actually,

$$\begin{aligned} m = \frac{1}{p}(x^2+1) &< \frac{1}{p}[(p-1)^2+1] \\ &< p. \end{aligned}$$

To prove (2), first note, as before, that m is odd. Suppose $m \neq 1$. Then $3 \leq m < p$. Choose y_1 and y_2 with

$$x_i \equiv y_i \pmod m, \; -\frac{m-1}{2} \leq y_i \leq \frac{m-1}{2}. \tag{3.26}$$

As in the proof of Theorem 3.11,

$$mn = y_1^2 + y_2^2, \qquad 1 \leq n < m. \tag{3.27}$$

Therefore multiplying (3.25) and (3.27), we get

$$m^2np = (x_1^2 + x_2^2)(y_1^2 + y_2^2) = (x_1y_1 + x_2y_2)^2 + (x_1y_2 - x_2y_1)^2. \quad (3.28)$$

Since $x_1 \equiv y_1 \pmod{m}$ and $y_2 \equiv x_2 \pmod{m}$, it is clear that

$$x_1y_2 \equiv x_2y_1 \pmod{m},$$

i.e., $m \mid x_1y_2 - x_2y_1$, and from the above equation, we see that $m \mid x_1y_1 + x_2y_2$. Thus

$$np = \left(\frac{x_1y_1 + x_2y_2}{m}\right)^2 + \left(\frac{x_1y_2 - x_2y_1}{m}\right)^2$$

contradicting the minimality of m. This completes the proof. □

Corollary 3.15. *An integer $n \geq 1$ is a sum of two squares if and only if no prime $p \equiv 3 \pmod 4$ occurs to an odd power in the factorization of n into powers of distinct primes.*

Proof. First suppose that $n = x^2 + y^2$. We will show that if p^α is the highest power of a prime $p \equiv 3 \pmod 4$ dividing n, then α is even. Suppose to the contrary that α is odd. If $d = (x, y)$, then $d^2 \mid n$ and

$$n_1 = x_1^2 + y_1^2, \quad (3.29)$$

where $x_1 = x/d$, $y_1 = y/d$, and $n_1 = n/d^2$. Clearly $(x_1, y_1) = 1$, and therefore p can divide at most one of the x_1, y_1, and n_1. Since α is odd and d^2 can cancel only an even power of p in n, there is a positive power of p left in the factorization of n_1, and p does not divide either of x_1, y_1. Regarding (3.29) as an equation over $\mathbb{F}_p$, we get $-1 = (x_1/y_1)^2$, *i.e.*,

$$\left(\frac{-1}{p}\right) = 1.$$

But this is impossible, because $p \equiv 3 \pmod 4$ implies that

$$\left(\frac{-1}{p}\right) = (-1)^{(p-1)/2} = -1.$$

Thus α cannot be odd.

Conversely write $n = m^2p_1 \cdots p_r$, $p_j \equiv 1 \pmod 4$, $j = 1, \ldots, r$. Now it is obvious from Theorem 3.14 and the repeated application of the identity (3.28) that $p_1 \ldots p_r$ is a sum of two squares, say $x_0^2 + y_0^2$. Therefore

$$n = (mx_0)^2 + (my_0)^2. \quad □$$

A form $f(\mathbf{x})$ of degree >0 always represents zero, because $f(\mathbf{0}) = 0$, where $\mathbf{0} = (0, \ldots, 0)$.

DEFINITION 3.16. A form $f(\mathbf{x})$ in $\mathbb{Z}[x_1, \ldots, x_n]$ *represents zero nontrivially* if $f(\mathbf{a}) = 0$ for some nonzero vector $\mathbf{a}$ with integer coordinates.

The quadratic form $x_1^2 - x_2^2$ represents zero nontrivially, whereas $x_1^2 + x_2^2$ does not. Our next example is the quadratic form

$$ax^2 + by^2 + cz^2.$$

We may assume that a, b, c are square-free and coprime in pairs. Legendre was the first to state and prove the precise conditions under which it represents zero nontrivially. For a historical account see Ref. 10.

THEOREM 3.17 (*Legendre*). *Suppose* $a, b, c \in \mathbb{Z}$ and abc *is square-free and* $\neq 0$. *The quadratic form* $f(\mathbf{x}) = ax^2 + by^2 + cz^2$ *represents zero nontrivially if and only if*

1. a, b, c, *do not have the same sign*;
2. $-bc$, $-ca$, $-ab$ *are quadratic residues modulo* $|a|$, *modulo* $|b|$ *and modulo* $|c|$, *respectively.*

The proof given here is due to Mordell and Skolem (cf. Ref. 6 or 7). We need some lemmas.

LEMMA 3.18. *Suppose* $a > 1$ *is a square-free integer such that* -1 *is a quadratic residue modulo* a. *Then the binary quadratic form* $x^2 + y^2$ *represents* a *nontrivially.*

PROOF. Choose s such that $-1 \equiv s^2 \pmod{a}$. The set

$$\{u - vs \,|\, 0 \leq u, v \leq |\sqrt{a}|\}$$

has $(1 + [\sqrt{a}])^2 > a$ integers. Therefore for two distinct pairs u_1, v_1 and u_2, v_2, we must have

$$u_1 - v_1 s \equiv u_2 - v_2 s \pmod{a}.$$

If we put $x_1 = u_1 - u_2$, $y_1 = v_1 - v_2$, then

$$x_1 \equiv y_1 s \pmod{a}. \tag{3.30}$$

Note that

$$0 \leq |x_1|, |y_1| \leq \sqrt{a}. \tag{3.31}$$

and

$$\mathbf{x} = (x_1, y_1) \neq \mathbf{0}. \tag{3.32}$$

In view of (3.31) and (3.32),

$$0 < x_1^2 + y_1^2 < 2a. \tag{3.33}$$

Now from (3.30) we obtain

$$\begin{aligned} x_1^2 + y_1^2 &\equiv s^2 y_1^2 + y_1^2 \\ &= y_1^2(s^2+1) \\ &\equiv 0 \pmod{a}, \end{aligned}$$

i.e., $x_1^2 + y_1^2$ is a multiple of a. Using (3.33), we get

$$x_1^2 + y_1^2 = a.$$

LEMMA 3.19. *Suppose A, B, C are positive real numbers such that the product $ABC = m$ is an integer. Then any linear congruence equation*

$$\alpha x + \beta y + \gamma z \equiv 0 \pmod{m}$$

has a nontrivial solution (x_0, y_0, z_0) with $|x_0| \le A$, $|y_0| \le B$, $|z_0| \le C$.

PROOF. The set

$$\{(x, y, z) \in \mathbb{Z}^3 \mid 0 \le x \le [A],\ 0 \le y \le [B],\ 0 \le z \le [C]\}$$

has $(1+[A])(1+[B])(1+[C]) > ABC = m$ elements. Therefore for two distinct triplets (x_i, y_i, z_i), $i = 1, 2$, we must have

$$\alpha x_1 + \beta y_1 + \gamma z_1 \equiv \alpha x_2 + \beta y_2 + \gamma z_2 \pmod{m}$$

and

$$x_0 = x_1 - x_2, \qquad y_0 = y_1 - y_2, \qquad z_0 = z_1 - z_2$$

is a required solution. □

LEMMA 3.20. *Suppose the quadratic form $f(\mathbf{x}) = ax^2 + by^2 + cz^2$ factors into linear factors modulo m_1 as well as modulo m_2. If m_1 and m_2 are coprime, then $f(\mathbf{x})$ factors into linear factors modulo $m_1 m_2$.*

PROOF. We are given that

$$f(\mathbf{x}) \equiv (\alpha_1 x + \beta_1 y + \gamma_1 z)(\alpha_1' x + \beta_1' y + \gamma_1' z) \pmod{m_1}$$

and

$$f(\mathbf{x}) \equiv (\alpha_2 x + \beta_2 y + \gamma_2 z)(\alpha_2' x + \beta_2' y + \gamma_2' z) \pmod{m_2}.$$

By the Chinese remainder theorem, choose α, β, γ and α', β', γ' such that

$$\alpha \equiv \alpha_i, \qquad \beta \equiv \beta_i, \qquad \gamma \equiv \gamma_i \qquad \pmod{m_i}$$

and

$$\alpha' \equiv \alpha_i', \qquad \beta' \equiv \beta_i', \qquad \gamma' \equiv \gamma_i' \qquad \pmod{m_i},$$

for $i = 1, 2$. Then the congruence

$$f(\mathbf{x}) \equiv (\alpha x + \beta y + \gamma z)(\alpha' x + \beta' y + \gamma' z)$$

holds modulo m_1 as well as modulo m_2. Since m_1 and m_2 are coprime, it must also hold modulo $m_1 m_2$. □

Proof of Theorem 3.17. First suppose that $f(\mathbf{x})$ represents zero nontrivially. Clearly a, b, c cannot all have the same sign. If (x, y, z) is a nontrivial solution of

$$ax^2 + by^2 + cz^2 = 0, \tag{3.34}$$

we may assume that x, y, z are coprime in pairs. To prove that $-bc$ is a quadratic residue modulo $|a|$, we first show that a, z are coprime. Suppose not and let $p \mid (a, z)$. Then $p \mid by^2$. But $(a, b) = 1$, so $p \mid y^2$, which implies that y, z are not coprime, a contradiction. Now choose u with $uz \equiv 1 \pmod{|a|}$. Because

$$\begin{aligned} ax^2 + by^2 + cz^2 &\equiv by^2 + cz^2 \\ &\equiv 0 \pmod{|a|}, \end{aligned} \tag{3.35}$$

multiplying each side of (3.35) by bu^2, we have

$$\begin{aligned} b^2u^2y^2 &\equiv -bcu^2z^2 \\ &\equiv -bc \pmod{|a|}. \end{aligned}$$

This shows that $-bc$ is a quadratic residue $\pmod{|a|}$. Similarly $-ca$, $-ab$ are quadratic residues modulo $|b|$ and modulo $|c|$, respectively.

Conversely, suppose (1) and (2) hold. If we change the signs of all of a, b, c, then (1), (2), and (3.34) still hold. Therefore, if necessary, by rearranging the variables we may assume that $a > 0$ and $b, c < 0$.

Because of our assumptions on a, b, c we choose r and c_1 such that $r^2 \equiv -bc \pmod{a}$ and $cc_1 \equiv 1 \pmod{a}$. Then

$$\begin{aligned} by^2 + cz^2 &\equiv cc_1(by^2 + cz^2) \\ &= c_1(bcy^2 + c^2z^2) \\ &\equiv c_1(c^2z^2 - r^2y^2) \\ &= c_1(cz + ry)(cz - ry) \\ &\equiv (z + c_1ry)(cz - ry) \pmod{a}, \end{aligned}$$

which shows that

$$\begin{aligned} f(\mathbf{x}) &= ax^2 + by^2 + cz^2 \\ &\equiv (z + c_1ry)(cz - ry) \pmod{a}, \end{aligned}$$

i.e., $f(\mathbf{x})$ factors into linear factors $\pmod{a}$. Similarly $f(\mathbf{x})$ factors into linear factors modulo $|b|$ as well as modulo $|c|$. By Lemma 3.20,

$$f(\mathbf{x}) \equiv (\alpha x + \beta y + \gamma z)(\alpha' x + \beta' y + \gamma' z) \pmod{abc}. \tag{3.36}$$

In Lemma 3.19, if we put $A = \sqrt{bc}$, $B = \sqrt{-ca}$, $C = \sqrt{-ab}$, we get a nontrivial solution (x_0, y_0, z_0) of

$$\alpha x + \beta y + \gamma z \equiv 0 \pmod{abc}, \tag{3.37}$$

with $|x_0| \le A$, $|y_0| \le B$, $|z_0| \le C$; i.e.,

$$x_0^2 \le bc, \qquad y_0^2 \le -ca, \qquad z_0^2 \le -ab.$$

Since bc is square-free, $x_0^2 = bc$ is possible only when $b = c = -1$. Similarly $y_0^2 = -ca$ (respectively $z_0^2 = -ab$) is possible only if $a = 1$, $c = -1$, (respectively $a = 1$, $b = -1$). Because $a > 0$ and $b, c < 0$, unless $b = c = -1$, we must have

$$ax_0^2 + by_0^2 + cz_0^2 \le ax_0^2 < abc$$

and (unless $a = 1$ also, in which case there is nothing to prove)

$$\begin{aligned} ax_0^2 + by_0^2 + cz_0^2 \ge by_0^2 + cz_0^2 &> b(-ac) + c(-ab) \\ &= -2abc. \end{aligned}$$

Thus except in the special case $b = c = -1$, we have the inequalities

$$-2abc < ax_0^2 + by_0^2 + cz_0^2 < abc. \tag{3.38}$$

Since (x_0, y_0, z_0) is a solution of (3.37), it is a solution of (3.36) also. Hence from (3.38), either

$$ax_0^2 + by_0^2 + cz_0^2 = 0,$$

in which case $f(\mathbf{x})$ represents zero nontrivially, or

$$ax_0^2 + by_0^2 + cz_0^2 = -abc,$$

in which case $x = -by_0 + z_0x_0$, $y = ax_0 + y_0z_0$, $z = z_0^2 + ab$ can be checked to be a nontrivial solution of $f(\mathbf{x}) = 0$.

In the special case of $b = c = -1$ and $a > 0$, by Lemma 3.18, $a = x_1^2 + y_1^2$ with $(x_1, y_1) \ne (0, 0)$. This gives a nontrivial solution $(1, y_1, z_1)$ of $f(\mathbf{x}) = 0$. □

3.4. Equivalence of Quadratic Forms

Let us consider, as an example, the quadratic form

$$f(\mathbf{x}) = 5x_1^2 + 16x_1x_2 + 13x_2^2.$$

To study what integers are represented by $f(\mathbf{x})$, we note that the substitution

$$\begin{aligned} x_1 &= 2y_1 - 3y_2, \\ x_2 &= -y_1 + 2y_2 \end{aligned} \tag{3.39}$$

transforms $f(\mathbf{x})$ to $g(\mathbf{y}) = y_1^2 + y_2^2$. Conversely,

$$\begin{aligned} y_1 &= 2x_1 + 3x^2, \\ y_2 &= x_1 + 2x_2 \end{aligned} \tag{3.40}$$

takes $g(\mathbf{y})$ back to $f(\mathbf{x})$.

Since the substitutions (3.39) and (3.40) are inverses of each other and set up a one-to-one correspondence between the integral vectors $\mathbf{x} = (x_1, x_2)$ and the integral vectors $\mathbf{y} = (y_1, y_2)$, it is clear that $f(\mathbf{x})$ and $g(\mathbf{x})$ represent the same set of integers. So we can answer our question by Corollary 3.15. Thus it is reasonable not to distinguish the two quadratic forms $f(\mathbf{x})$ and $g(\mathbf{x})$.

In matrix notations, we can write

$$\begin{aligned} f(\mathbf{x}) &= 5x_1^2 + 16x_1x_2 + 13x_2^2 \\ &= (x_1 \quad x_2)\begin{pmatrix} 5 & 8 \\ 8 & 13 \end{pmatrix}\begin{pmatrix} x_1 \\ x_2 \end{pmatrix} \\ &= \mathbf{x}A\mathbf{x}' \\ &\overset{\text{def}}{=} A[\mathbf{x}], \end{aligned}$$

where

$$A = \begin{pmatrix} 5 & 8 \\ 8 & 13 \end{pmatrix}, \qquad \mathbf{x} = (x_1 \quad x_2)$$

and $\mathbf{x}'$ denotes the transpose of a matrix $\mathbf{x}$.

The substitutions (3.39) and (3.40) can be written as $\mathbf{x} = \mathbf{y}T^{-1}$ and $\mathbf{y} = \mathbf{x}T$, respectively, where

$$T = \begin{pmatrix} 2 & 1 \\ 3 & 2 \end{pmatrix}$$

and

$$T^{-1} = \begin{pmatrix} 2 & -1 \\ -3 & 2 \end{pmatrix}$$

is the inverse of T. Moreover,

$$\begin{aligned} g(\mathbf{y}) &= I[\mathbf{x}T] \\ &= (I[T])[\mathbf{x}] \\ &= A[\mathbf{x}] \end{aligned}$$

and

$$\begin{aligned} f(\mathbf{x}) &= A[\mathbf{y}T^{-1}] \\ &= A[T^{-1}][\mathbf{y}] \\ &= I[\mathbf{y}], \end{aligned}$$

I being a 2×2 identity matrix. With this motivation we pass on to the general case.

A quadratic form

$$f(\mathbf{x}) = f(x_1, \ldots, x_n) = \sum_{\substack{i,j=1 \\ j \geq i}}^{n} b_{ij} x_i x_j$$

over $\mathbb{R}$ can be written as

$$f(\mathbf{x}) = \mathbf{x} A \mathbf{x}' \stackrel{\text{def}}{=} A[\mathbf{x}],$$

where (the symmetric matrix) $A = (a_{ij})$ is the *matrix of the quadratic form* $f(\mathbf{x})$ defined by

$$a_{ij} = \begin{cases} b_{ij}, & \text{if } i = j; \\ a_{ji} = \frac{1}{2} b_{ij}, & \text{if } j > i. \end{cases}$$

We sometimes say that the quadratic form $f(\mathbf{x})$ is *represented by* the matrix A. Conversely, any symmetric matrix A defines a quadratic form $f(\mathbf{x}) = A[x]$. Two quadratic forms $f(\mathbf{x})$ and $g(\mathbf{x})$ represented by the matrices A and B, respectively, are *equivalent* if there is a unimodular substitution that takes $f(\mathbf{x})$ into $g(\mathbf{x})$. This means that there is a *unimodular* matrix, i.e., a matrix U in $M(n, \mathbb{Z})$ with $\det(U) = \pm 1$, such that $f(\mathbf{x}U) = g(\mathbf{x})$. In terms of matrices this is equivalent to $B = A[U]$.

If

$$\mathbb{Z}^n = \{(x_1, \ldots, x_n) \mid x_i \in \mathbb{Z},\ i = 1, \ldots, n\},$$

then the map

$$m_U \colon \mathbb{Z}^n \to \mathbb{Z}^n$$

given by $m_U \mathbf{x} = \mathbf{x} U$ is a bijection, so $f(\mathbf{x})$ and $g(\mathbf{x})$ represent the same set of integers. Moreover, the determinant is invariant under the map $A \to A[U]$. We may define the determinant of an equivalence class of quadratic forms as $|A|$, the determinant of the matrix A of any quadratic form taken from this equivalence class.

3.5. Minima of Positive Quadratic Forms

All the forms considered in the rest of this chapter are over $\mathbb{Z}$, unless stated otherwise. The case of $n = 1$, i.e., a quadratic form in one variable, is not very interesting, so we shall also assume that $n > 1$.

Definition 3.21. A quadratic form $f(\mathbf{x})$ with matrix A is *positive* (*definite*), written as $f > 0$ or $A > 0$, if $f(\mathbf{x}) > 0$ for all $\mathbf{x} \neq \mathbf{0}$, in

$$\mathbb{R}^n = \{(x_1, \ldots, x_n) \mid x_i \in \mathbb{R},\ i = 1, \ldots, n\}.$$

An example of a positive form is

$$f(\mathbf{x}) = x_1^2 + \cdots + x_n^2.$$

If $f(\mathbf{x})$ is positive and $g(\mathbf{x})$ is equivalent to $f(\mathbf{x})$, then $g(\mathbf{x})$ is also positive. Among the nonzero integers represented by $f(\mathbf{x})$, there is the smallest one denoted by $\mu(f)$ or $\mu(A)$, i.e.,

$$\mu(f) = \mu(A) = \min\{f(\mathbf{x}) \mid \mathbf{x} \in \mathbb{Z}^n, \mathbf{x} \neq 0\}.$$

Clearly $\mu(f) > 0$. If $f(\mathbf{x})$ is equivalent to $g(\mathbf{x})$, then $\mu(f) = \mu(g)$. To study $\mu(f)$, we need the following theorem.

Theorem 3.22. *Suppose* $\mathbf{u} = (x_1, \ldots, x_n)$ *is a nonzero vector in* $\mathbb{Z}^n$. *If* $d = \text{g.c.d.}(x_1, \ldots, x_n)$ *then there is a matrix* A *in* $M(n, \mathbb{Z})$, *of determinant* d, *whose first row is* $\mathbf{u}$.

Proof. The proof is by induction on n. If $n = 2$, then the $\text{g.c.d.}(x_1, x_2) = d = \alpha x_1 + \beta x_2$ for some α, β in $\mathbb{Z}$. We put

$$A = \begin{pmatrix} x_1 & x_2 \\ -\beta & \alpha \end{pmatrix}.$$

Now suppose that $n > 2$. Let $c = \text{g.c.d.}(x_1, \ldots, x_{n-1})$. Then $d = \text{g.c.d.}(c, x_n)$ and therefore, for some r, s in $\mathbb{Z}$, we have

$$rc - sx_n = d. \tag{3.41}$$

By induction hypothesis, there is a matrix C in $M(n-1, \mathbb{Z})$ of determinant c and with the vector $(x_1, \ldots, x_{n-1})$ as its first row. If we put

$$A = \left(\begin{array}{ccc|c} & & & x_n \\ & & & 0 \\ & C & & \vdots \\ & & & 0 \\ \hline sx_1/c & \cdots & sx_{n-1}/c & r \end{array}\right),$$

then A is in $M(n, \mathbb{Z})$ and $\mathbf{u}$ is the first row of A. Expanding the determinant $|A|$ of A by its last column, we get

$$|A| = rc + (-1)^{n-1} x_n |B|, \tag{3.42}$$

where the matrix B is obtained from C by multiplying its first row by s/c and then interchanging it successively with the following rows until it becomes the last row. Therefore,

$$|B| = (-1)^{n-2}(s/c)|C| = (-1)^{n-2}s.$$

Substituting it in (3.42) and using (3.41), we get

$$\begin{aligned}|A| &= rc - sx_n \\ &= d.\end{aligned}$$

□

COROLLARY 3.23. *If* $\mathbf{u} = (x_1, \ldots, x_n) \in \mathbb{Z}^n$ *is primitive, i.e., with coprime coordinates* $x_1, \ldots, x_n$, *then* $\mathbf{u}$ *is the first row of a unimodular matrix* U.

COROLLARY 3.24. *If* $x_1, \ldots, x_n \in \mathbb{Z}$ *with* $\text{g.c.d.}(x_1, \ldots, x_n) = d$, *then there exist integers* $\lambda_1, \ldots, \lambda_n$ *such that*

$$\lambda_1 x_1 + \cdots + \lambda_n x_n = d.$$

PROOF. If A is the matrix given by Theorem 3.22, we expand $|A|$ using the first row of A. □

LEMMA 3.25. *Suppose A is positive and $\mu(A) = m$. Then there is a unimodular matrix U such that if $B = (b_{ij}) = A[U]$, then $b_{11} = m$.*

PROOF. Suppose $\mu(A) = A[\mathbf{x}]$ for $\mathbf{x} = (x_1, \ldots, x_n)$ in $\mathbb{Z}^n$. Then clearly $x_1, \ldots, x_n$ are coprime, because if d is a common divisor of $x_1, \ldots, x_n$ and $x_i = da_i$, then letting $\mathbf{a} = (a_1, \ldots, a_n)$, we get

$$\mu(A) = A[\mathbf{x}] = A[d\mathbf{a}] = d^2 A[\mathbf{a}],$$

and the minimality of $A[\mathbf{x}]$ shows that $d^2 = 1$. Thus the minimum is attained at a primitive vector $\mathbf{x}$. If $\mathbf{x}$ is the first row of a unimodular matrix U, then U has the required properties. □

Suppose A, B are two symmetric positive matrices in $M(n, \mathbb{Z})$ such that $B = tA$ for some $t \in \mathbb{Z}$. Clearly $t > 0$ and

$$\mu(B) = \mu(tA) = t\mu(A). \tag{3.43}$$

On the other hand, the determinant

$$|B| = |tA| = t^n|A|,$$

so that

$$|B|^{1/n} = t|A|^{1/n}. \tag{3.44}$$

(Note that the determinant of a positive quadratic form is positive.) Thus, in view of (3.43) and (3.44), it seems natural to compare $\mu(A)$ and $|A|^{1/n}$. In this regard we have a well-known result of Hermite.

THEOREM 3.26 (*Hermite*). *If A is an $n \times n$ matrix of a positive quadratic form, then*

$$\mu(A) \le (\tfrac{4}{3})^{(n-1)/2} |A|^{1/n}. \tag{3.45}$$

PROOF. Our proof is as in Siegel [9]. We use induction on n. For $n = 1$, $A = (a)$ with $a > 0$. Clearly $\mu(A) = a$ and $|A| = a$ and (3.45) is a triviality.

Suppose the theorem is true for $n - 1$ ($n > 1$). We will show that it is true for n also. Let $A = (a_{ij})$ be $n \times n$ and $\mu(A) = m$. By Lemma 3.25, choose a unimodular matrix U such that

$$B = A[U] = \begin{pmatrix} m & * \\ * & * \end{pmatrix}.$$

Since $\mu(A) = \mu(B)$ and $|A| = |B|$, if necessary we replace A by B to assume that $a_{11} = m$.

Partitioning A as

$$A = \begin{pmatrix} m & b \\ b' & A_1 \end{pmatrix},$$

we see that

$$A = \begin{pmatrix} m & 0 \\ 0 & W \end{pmatrix} \left[\begin{pmatrix} 1 & 0 \\ m^{-1}b' & I \end{pmatrix} \right],$$

with $W = A_1 - m^{-1}b'b$. (I is the $n - 1 \times n - 1$ identity matrix.) Also

$$|A| = m|W|. \tag{3.46}$$

If $\mathbf{x} = (x_1, \mathbf{y}) \in \mathbb{Z}^n$ with $\mathbf{y} \in \mathbb{Z}^{n-1}$, then

$$A[\mathbf{x}] = m(x_1 + m^{-1}\mathbf{y}b')^2 + W[\mathbf{y}]. \tag{3.47}$$

Now choose x_1 such that

$$|x_1 + m^{-1}\mathbf{y}b'| \le \tfrac{1}{2}, \tag{3.48}$$

and since $W > 0$, choose $\mathbf{y}$ so that $W[\mathbf{y}] = \mu(W)$. Then by (3.47), (3.48) and the induction hypothesis,

$$\begin{aligned} m = \mu(A) &\le A[\mathbf{x}] \\ &\le \tfrac{1}{4}m + (\tfrac{4}{3})^{(n-2)/2} |W|^{1/(n-1)}. \end{aligned}$$

Substituting the value of $|W|$ from (3.46), this yields

$$\mu(A) \le (\tfrac{4}{3})^{(n-1)/2} |A|^{1/n}. \qquad \square$$

3.6. Reduction of Positive Quadratic Forms

We shall discuss only the special case of *binary* quadratic forms, i.e., quadratic forms in two variables. For the general case, see Ref. 9. For a historical account, Ref. 10 may be consulted.

A binary quadratic form over $\mathbb{Z}$ can be written as

$$f(x, y) = ax^2 + bxy + cy^2. \tag{3.49}$$

THEOREM 3.27. *A binary quadratic form* (3.49) *is positive if and only if* $a > 0$, $c > 0$ *and its discriminant* $D = b^2 - 4ac < 0$.

PROOF. First suppose that $f > 0$. Then $f(1, 0) = a > 0$ and $f(0, 1) = c > 0$. Writing

$$f(x, y) = \frac{1}{4a}[(2ax + by)^2 + (4ac - b^2)y^2], \tag{3.50}$$

we see that $f(-b, 2a) = a(4ac - b^2) > 0$, which proves that $b^2 - 4ac < 0$.

Conversely, if $a > 0$, $c > 0$, and $b^2 - 4ac < 0$, then by (3.50),

$$f(x, y) > 0 \qquad \text{if } y \neq 0.$$

If $y = 0$, then for $x \neq 0$,

$$f(x, 0) = ax^2 > 0. \qquad \square$$

COROLLARY 3.28. *The number of representations of a given integer* $m > 0$ *by a positive quadratic form* (3.49) *is finite*, i.e.,

$$f(x, y) = m$$

for only finitely many integer vectors (x, y).

PROOF. Note that by (3.50),

$$|y| \leq 2(-am/D)^{1/2},$$

where $D = b^2 - 4ac$. Thus y can have only finitely many integer values. For each such y, there are at most two values of x. $\square$

COROLLARY 3.29. *The discriminant of a quadratic form is invariant under unimodular substitutions.*

PROOF. The discriminant $D = b^2 - 4ac$ of (3.49) is related to the determinant of its matrix

$$A = \begin{pmatrix} a & b/2 \\ b/2 & c \end{pmatrix}$$

by $D = -4|A|$. Under the unimodular substitutions $\mathbf{x} \to \mathbf{x}U$, the matrix A is taken to $A[U]$. But $|A| = |A[U]|$. □

DEFINITION 3.30. A positive binary quadratic form (3.49) is *reduced* if $c \geq a \geq b \geq 0$.

Note that $a > 0$; otherwise $f(x, y)$ reduces to a form in one variable.

THEOREM 3.31. *Any positive binary quadratic form is equivalent to one and only one reduced form.*

PROOF. First we show that any positive binary quadratic form

$$f(x, y) = Ax^2 + Bxy + Cy^2$$

is equivalent to a reduced form. It is enough to show that it is equivalent to $ax^2 + bxy + cy^2$ with $c \geq a \geq |b| \geq 0$, for if b is negative, the unimodular substitution $x = X$, $y = -Y$ will take $ax^2 + bxy + cy^2$ to a reduced form.

Suppose $a = \mu(f)$. By Lemma 3.25, $f(x, y)$ is equivalent to a positive binary quadratic form $ax^2 + b_1xy + c_1y^2$, and then the unimodular substitution $x \to x - ky$, $y \to y$ takes $ax^2 + b_1xy + c_1y^2$ to $ax^2 + (b_1 - 2ak)xy + cy^2$. Choosing k such that

$$\left| k - \frac{b_1}{2a} \right| \leq \frac{1}{2},$$

i.e., $|b_1 - 2ak| \leq a$ and letting $b = b_1 - 2ak$, we arrive at a required positive binary quadratic form $ax^2 + bxy + cy^2$ such that

$$c = f(0, 1) \geq \mu(f) = a \geq |b| \geq 0.$$

To complete the proof we must show that any two reduced positive binary quadratic forms

$$f(x, y) = ax^2 + bxy + cy^2$$

and

$$g(X, Y) = AX^2 + BXY + CY^2$$

that are equivalent have to be identical, i.e., $a = A$, $b = B$, $c = C$.

First we show that $a = A$. To do this we show that $a \geq A$ implies that $A \geq a$. Let the unimodular substitution taking the first form into the second be

$$\begin{pmatrix} x \\ y \end{pmatrix} = \begin{pmatrix} \alpha & \beta \\ \gamma & \delta \end{pmatrix} \begin{pmatrix} X \\ Y \end{pmatrix}.$$

Then

$$A = a\alpha^2 + b\alpha\gamma + c\gamma^2,$$
$$B = 2a\alpha\beta + b(\alpha\delta + \beta\gamma) + 2a\gamma\delta, \tag{3.51}$$
$$C = a\beta^2 + b\beta\delta + c\delta^2.$$

Since $c \geq a \geq b \geq 0$ and $\alpha^2 + \gamma^2 \geq 2|\alpha\gamma|$, we have

$$\begin{aligned} A &= a\alpha^2 + b\alpha\gamma + c\gamma^2 \geq a\alpha^2 + c\gamma^2 - b|\alpha\gamma| \\ &\geq a\alpha^2 + a\gamma^2 - b|\alpha\gamma| \geq 2a|\alpha\gamma| - b|\alpha\gamma| \\ &\geq a|\alpha\gamma|. \end{aligned} \tag{3.52}$$

Now $a \geq A$ shows that $|\alpha\gamma| \leq 1$. α and γ cannot both be zero, so if $|\alpha\gamma| = 0$,

$$A \geq a\alpha^2 + c\gamma^2 \geq a, \tag{3.53}$$

and if $|\alpha\gamma| = 1$, it is immediate from (3.52) that $A \geq a$.

By Corollary 3.29, $b^2 - 4ac = B^2 - 4AC$. Since $a = A > 0$, it suffices to prove only either $b = B$ or $c = C$. Suppose $c \neq C$. Without loss of generality, let $c > C \geq A = a > 0$. Then $|\alpha\gamma| = 0$, because $|\alpha\gamma| = 1$ would imply $c\gamma^2 > a\gamma^2$, giving a strict inequality $A > a$ in (3.52). Actually, $\gamma = 0$, because if $\gamma \neq 0$, then $c > a$ and (3.53) imply that $A > a$.

Now from $\gamma = 0$ and $\alpha\delta - \beta\gamma = \pm 1$, we get $\alpha\delta = \pm 1$, and by (3.51), $B = 2a\alpha\beta \pm b$. There are two cases:

1. $B = 2a\alpha\beta + b$. Since $0 \leq b \leq a$ and $0 \leq B \leq A = a$, we must have $|B - b| \leq a$. Since $B - b$ is a multiple of $2a$, $B - b = 0$, i.e., $B = b$.

2. $B = 2a\alpha\beta - b$. Now $0 \leq B + b \leq 2a$ and $B + b = 2a\alpha\beta$, so either $B + b = 0$ or $B + b = 2a$. But $0 \leq b \leq a$, $0 \leq B \leq A = a$, so if $B + b = 0$, then $B = b = 0$ and if $B + b = 2a$, then we must have $B = b = a$.

In any case, $B = b$ and the theorem is proved. □

Corollary 3.32. *There are only finitely many inequivalent positive binary quadratic forms of a given discriminant.*

Proof. By Theorem 3.27, there are no positive binary quadratic forms of positive discriminant. So it is enough to show that there are only finitely many reduced positive binary quadratic forms

$$ax^2 + bxy + cy^2$$

of discriminant $D = b^2 - 4ac < 0$. Since $c \geq a \geq b \geq 0$, $-D = 4ac - b^2 \geq 3ac \geq 3a^2$ or $a \leq (-D/3)^{1/2}$. Thus a and b can have only finitely many integer values. Since c is related to a, b by $c = (b^2 - D)/4a$, for each pair a, b, there is at most one value of c in $\mathbb{Z}$. □

For further discussion, see a recent paper of D. Goldfeld [2].

EXERCISES.

1. Find the reduced form equivalent to $3x^2 + 7xy + 5y^2$.
2. Find all the reduced positive forms of discriminant $D \geq -20$.

References

1. W. J. Ellison, Waring's problem, *Am. Math. Mon.*, **78**, 10–36 (1971).
2. D. Goldfeld, Gauss's class number problem for imaginary quadratic fields, *Bull. Am. Math. Soc.* **13**, 23–37 (1985).
3. D. Hilbert, Beweis für die Darstellbarkeit der ganzen Zahlen durch eine feste Anzahl n-ter Potenzen (Waringsches Problem), *Math. Ann.* **67**, 281–300 (1909).
4. J.-I. Igusa, *Lectures on Forms of Higher Degree* (Notes by S. Raghavan), Tata Institute of Fundamental Research, Bombay (1978).
5. Yu. I. Manin, *Cubic Forms—Algebra, Geometry, Arithmetic* (translated from Russian by M. Hazewinkel), North-Holland, Amsterdam (1974).
6. L. J. Mordell, *Diophantine Equations*, Academic, London (1969).
7. I. Niven and H. S. Zuckerman, *An Introduction to the Theory of Numbers*, Wiley, New York (1980).
8. H. Pieper, *Variationen über ein zahlentheoretisches Thema von C. F. Gauss*, Birkhäuser, Basel (1978).
9. C. L. Siegel, *Lectures on Quadratic Forms* (Notes by K. G. Ramanathan), Tata Institute of Fundamental Research, Bombay (1957).
10. A. Weil, *Number Theory—An Approach through History*, Birkhäuser, Boston (1984).

4

Algebraic Number Fields

4.1. Introduction

Let us consider the diophantine equation

$$x^2 - dy^2 = 1, \tag{4.1}$$

erroneously called Pell's equation. (For its history, see Ref. 9.) Here $d \neq 0$ is a square-free integer. We seek the integer solutions of (4.1). If $d < 0$, these solutions are $(\pm 1, 0)$ for $d < -1$ and $(\pm 1, 0)$, $(0, \pm 1)$ for $d = -1$. However, if $d > 1$, it is a nontrivial fact that (4.1) has infinitely many solutions in integers. If we let G denote the set of these solutions, then G has a group structure (cf. Exercise 2.4). Moreover, up to multiplication by -1 [i.e., $-(x, y) = (-x, -y)$], G is an infinite cyclic group. A generator is a solution with the smallest $|y_1|$ (and hence the smallest $|x_1|$) > 0.

To prove this we look at this problem algebraically as follows. For reasons to be clarified later, let us restrict ourselves to the case of $d \equiv 2, 3 \pmod 4$. It is easy to check that

$$K = \mathbb{Q}(\sqrt{d}) = \{r + s\sqrt{d} \mid r, s \in \mathbb{Q}\}$$

is a subfield of $\mathbb{C}$. (A subset K of a field L is a *subfield* of L if $1 \in K$, K is a subring of L and for each nonzero x in K, $x^{-1} \in K$.) In fact, one has only to check that for $\alpha = r + s\sqrt{d} \neq 0$, $(1/\alpha) = r_1 + s_1\sqrt{d}$ for some r_1, s_1 in $\mathbb{Q}$. Note that K can be considered as a vector space of dimension two over the field of scalars $\mathbb{Q}$. The set

$$A = \mathbb{Z}[\sqrt{d}] = \{x + y\sqrt{d} \mid x, y \in \mathbb{Z}\}$$

is a subring of K.

For $\alpha = x + y\sqrt{d}$ in A, put $\bar{\alpha} = x - y\sqrt{d}$ and call it the conjugate of α. The norm function $N\colon A \to \mathbb{Z}$ is defined by

$$N(\alpha) = \alpha\bar{\alpha}.$$

Another important function is the trace Tr: $A \to \mathbb{Z}$, defined by

$$\mathrm{Tr}(\alpha) = \alpha + \bar{\alpha}.$$

Any α in A is a root of a monic polynomial $f(X)$ in $\mathbb{Z}[X]$ (monic means the leading coefficient is 1). In fact,

$$\begin{aligned} f(X) &= (X - \alpha)(X - \bar{\alpha}) \\ &= X^2 - \mathrm{Tr}(\alpha)X + N(\alpha). \end{aligned}$$

Conversely, as we shall see later, any α in K that is a root of a monic polynomial over $\mathbb{Z}$ is in A. Thus A is precisely the ring of integers of K (i.e., the roots in K of monic polynomials over $\mathbb{Z}$).

The *group of units* $R^\times$ of a ring R with identity 1 is by definition the group of invertible elements of R, i.e.,

$$R^\times = \{u \in A \mid uv = vu = 1 \text{ for some } v \text{ in } R\}.$$

The group $A^\times$ can be characterized as

$$A^\times = \{u = x + y\sqrt{d} \in A \mid N(u) = x^2 - dy^2 = \pm 1\}.$$

If A has no element of norm -1, then G is isomorphic to $A^\times$ (otherwise, to $A^{\times 2} = \{u^2 \mid u \in A^\times\}$), so that the group G of integer solutions of (4.1) is completely determined by $A^\times$.

In general, suppose K is a subfield of $\mathbb{C}$. Then K contains 1 and hence $\mathbb{Z}$. Consequently $\mathbb{Q} \subseteq K$. Suppose K considered as a vector space over $\mathbb{Q}$ is finite dimensional and A is its ring of integers (roots in K of monic polynomial over $\mathbb{Z}$).

What is the structure of the group $A^\times$? This chapter is devoted to proving a famous theorem of Dirichlet that answers this question.

4.2. Number Fields

If k is a subfield of a field K, we call K a *field extension* of k and write it as K/k. A field extension K/k is a *finite extension* if the dimension $[K: k]$ of K as a vector space over k is finite. We call $n = [K: k]$ the *degree of the extension K/k*. We say that K/k is a *quadratic extension* or a *cubic extension* according as $n = 2$ or 3.

DEFINITION 4.1. If $K/\mathbb{Q}$ is a finite extension, K is called an (*algebraic*) *number field.*

A number field K is called a *quadratic field* or a *cubic field* according as $[K: \mathbb{Q}] = 2$ or 3.

DEFINITION 4.2. A complex number α is called an *algebraic number* if $f(\alpha) = 0$ for a nonzero polynomial $f(x)$ in $\mathbb{Q}[x]$. Otherwise, we say that α is *transcendental.*

Some well-known transcendental numbers are e and π. For proof see Ref. 1, pp. 4–6.

If K/k is an extension of number fields with $n = [K: k]$ and α is a nonzero element of K, then the $n + 1$ vectors

$$1, \alpha, \ldots, \alpha^n$$

must be linearly dependent over k, so that for some scalars $a_0, a_1, \ldots, a_n$, not all zero, we must have

$$a_0 + a_1\alpha + \cdots + a_n\alpha^n = 0.$$

This shows that every element of K is a root of a polynomial over k of degree at most n. Suppose for $\alpha \neq 0$ in K, $f(x)$ is a polynomial over k of the smallest degree with $f(\alpha) = 0$. If $g(x)$ is another polynomial over k with $g(\alpha) = 0$, we write

$$g(x) = q(x)f(x) + r(x)$$

with $q(x)$, $r(x)$ in $k[x]$ and the degree of the remainder $r(x) < \deg f(x)$, to get $r(\alpha) = 0$. The minimality condition on $\deg f(x)$ implies that $r(x) = 0$, i.e., $g(x) = q(x)f(x)$. Thus by definition, $f(x)$ divides $g(x)$ in $k[x]$, which we write as $f(x)|g(x)$. We may also say that $g(x)$ is a multiple of $f(x)$ in $k[x]$. Moreover, if $\deg g(x) = \deg f(x)$, then $g(x) = cf(x)$, for some constant $c \neq 0$ in k. Thus if we require the leading coefficient of $f(x)$ to be 1, then $f(x)$ is unique.

DEFINITION 4.3. A polynomial $a_0 + a_1x + \cdots + a_nx^n$ over a (commutative) ring A with 1 is called *monic* if $a_n = 1$.

DEFINITION 4.4. The *minimal polynomial* over k of a nonzero algebraic number α is the monic polynomial $f(x)$ over k of the smallest degree such that $f(\alpha) = 0$.

Clearly the minimal polynomial $f(x)$ of α is *irreducible over* k, i.e., there are no two polynomials $f_1(x)$, $f_2(x)$ over k such that

1. $f(x) = f_1(x)f_2(x)$

and

2. $0 < \deg f_j(x) < \deg f(x), \qquad j = 1, 2.$

This is so because otherwise α will be a root of either $f_1(x)$ or $f_2(x)$, contradicting the definition of $f(x)$.

From now on, unless stated otherwise, $k = \mathbb{Q}$ and "minimal polynomial" will mean "minimal polynomial over $k = \mathbb{Q}$."

Suppose $f(x)$ is the minimal polynomial of an algebraic number $\alpha \neq 0$ and $\deg f(x) = n$. Then $f(x)$ has n roots $\alpha_1 = \alpha, \ldots, \alpha_n$ in $\mathbb{C}$ and

$$f(x) = (x - \alpha_1) \cdots (x - \alpha_n).$$

Moreover, $\alpha_1, \ldots, a_n$ are distinct. For otherwise, if say $\alpha_1 = \alpha_2$, then

$$f(x) = (x - \alpha)^2 q(x),$$

which on differentiating gives

$$f'(x) = 2(x - \alpha)q(x) + (x - \alpha)^2 q'(x),$$

showing that α is a root of the polynomial $f'(x)$ of degree smaller than $\deg f(x)$.

DEFINITION 4.5. Suppose $f(x)$ is the minimal polynomial of an algebraic number $\alpha \neq 0$ and $\alpha_1 = \alpha, \ldots, \alpha_n$ are the distinct roots of $f(x)$. Then $\alpha_1, \ldots, \alpha_n$ are called the *conjugates* of α.

Note the symmetry. The minimal polynomial of α is also the minimal polynomial of its conjugates, so that α_i, α_j are conjugates of each other for all $i, j = 1, \ldots, n$.

For a given set $\beta_1, \ldots, \beta_n$ of complex numbers, not necessarily algebraic, let $\mathbb{Q}(\beta_1, \ldots, \beta_n)$ denote the field obtained from $\mathbb{Q}$ by adjoining $\beta_1, \ldots, \beta_n$, i.e., the smallest field containing $\mathbb{Q}$ as well as $\beta_1, \ldots, \beta_n$. It is the intersection of all subfields of $\mathbb{C}$ that contain $\mathbb{Q}$ and $\beta_1, \ldots, \beta_n$. [In particular, being a vector space of finite dimension over $\mathbb{Q}$, any number field $K = \mathbb{Q}(\alpha_1, \ldots, \alpha_n)$ for some algebraic numbers $\alpha_1, \ldots, \alpha_n$.] Clearly

$$\mathbb{Q}(\beta_1, \ldots, \beta_n) = \left\{ \frac{\phi_1(\beta_1, \ldots, \beta_n)}{\phi_2(\beta_1, \ldots, \beta_n)} \,\middle|\, \phi_1, \phi_2 \in \mathbb{Q}[x_1, \ldots, x_n] \text{ and } \phi_2(\beta_1, \ldots, \beta_n) \neq 0 \right\},$$

the *field of rational functions in* $\beta_1, \ldots, \beta_n$. In particular, suppose α is algebraic with minimal polynomial $f(x)$ of degree n. If $g(x)$ is another polynomial, as we have seen, $g(\alpha) = r(\alpha)$ for a polynomial $r(x)$ of degree $< n$. Therefore,

$$\mathbb{Q}(\alpha) = \left\{ \frac{f_1(\alpha)}{f_2(\alpha)} \,\middle|\, f_j(x) \in \mathbb{Q}[x], \deg f_j(x) < n \text{ and } f_2(\alpha) \neq 0 \right\}.$$

In fact, we shall now show that

$$\mathbb{Q}(\alpha) = \mathbb{Q}[\alpha] = \{g(\alpha) \mid g(x) \in \mathbb{Q}[x], \deg g(x) < n\}.$$

We need first a little bit of preparation.

4.3. Discriminant of a Polynomial

Throughout this section k will denote an arbitrary field. The *greatest common divisor* (g.c.d.) of two polynomials

$$\begin{aligned} f(x) &= a_0 + a_1 x + \cdots + a_n x^n, \\ g(x) &= b_0 + b_1 x + \cdots + b_m x^m \end{aligned} \tag{4.2}$$

of degree n, m respectively and defined over k is the monic polynomial $d(x) = (f(x), g(x))$ in $k[x]$ such that (1) $d(x)$ divides both $f(x)$, $g(x)$ (in the ring $k[x]$) and (2) if another polynomial $h(x)$ in $k[x]$ divides both $f(x)$ and $g(x)$, then $h(x)$ divides $d(x)$ also.

The existence of such a polynomial $d(x)$ can be shown via the division algorithm (see any book on algebra or imitate Theorem 1.22), and can be used to prove unique factorization for polynomials. The *resultant* or the *elimination* $R(f, g)$ of $f(x)$ and $g(x)$ is the following $(m+n) \times (m+n)$ determinant:

$$R(f, g) = \begin{vmatrix} a_0 & a_1 & \ldots & a_n & & & \\ & a_0 & a_1 & \ldots & a_n & & \\ & & & & \ldots & & \\ & & a_0 & a_1 & & \ldots & a_n \\ b_0 & b_1 & \ldots & b_m & & & \\ & b_0 & b_1 & \ldots & b_m & & \\ & & & & \ldots & & \\ & & b_0 & b_1 & & \ldots & b_m \end{vmatrix} \begin{matrix} \left.\vphantom{\begin{matrix}a\\a\\a\\a\end{matrix}}\right\} m \text{ rows} \\ \left.\vphantom{\begin{matrix}a\\a\\a\\a\end{matrix}}\right\} n \text{ rows} \end{matrix}$$

The missing entries are all zeros.

THEOREM 4.6. *Let* $d(x) = (f(x), g(x))$. *Then* $\deg d(x) > 0$ *if and only if* $R(f, g) = 0$.

For the proof we need the following lemma.

LEMMA 4.7. *Let* f, g, *and* d *be as above. Then* $\deg d(x) > 0$ *if and only if there are nonzero* $f_1, g_1 \in k[x]$, *such that* $\deg f_1 < \deg f$, $\deg g_1 < \deg g$ *and*

$$f_1 g = f g_1. \tag{4.3}$$

PROOF OF THE LEMMA. If $\deg d(x) > 0$, then $f = df_1$, $g = dg_1$ with $\deg f_1 < \deg f$, $\deg g_1 < \deg g$. Clearly, $fg_1 = f_1 g$.

Conversely, if (4.3) holds, then every irreducible factor of g appears in the factorization of fg_1. Since $\deg g_1 < \deg g$, some irreducible factor of g must divide f and hence $\deg d(x) > 0$. □

PROOF OF THEOREM 4.6. By Lemma 4.7, it is enough to show that (4.3) holds if and only if $R(f, g) = 0$. Let

$$f_1(x) = \sum_{j=1}^{n} \alpha_j x^{j-1}, \qquad g_1(x) = \sum_{j=1}^{m} \beta_j x^{j-1}.$$

It is obvious (by comparing the coefficients) that (4.3) is equivalent to the existence of a nonzero solution $(\beta_1, \ldots, \beta_m; \alpha_1, \ldots, \alpha_n)$ of the following system of linear equations:

$$\begin{aligned}
a_0\beta_1 &= b_0\alpha_1, \\
a_1\beta_1 + a_0\beta_2 &= b_1\alpha_1 + b_0\alpha_2, \\
&\cdots \\
&\cdots \\
a_n\beta_m &= b_m\alpha_n,
\end{aligned}$$

which is equivalent to the vanishing of the determinant

$$\begin{vmatrix}
a_0 & & \cdots & b_0 & & \cdots \\
a_1 & a_0 & \cdots & b_1 & b_0 & \cdots \\
 & a_1 & \cdots & & b_1 & \cdots \\
\vdots & \vdots & & \vdots & \vdots &
\end{vmatrix}.$$

This determinant is just the transpose of (and thus equal to) $R(f, g)$. □

DEFINITION 4.8. If $f(x) = a_0 + a_1 x + \cdots + a_n x^n$ $(a_n \neq 0)$ is a polynomial in $k[x]$, the *discriminant* $\Delta(f)$ of $f(x)$ is

$$\Delta(f) = (-1)^{n(n-1)/2} \frac{1}{a_n} R(f, f')$$

EXERCISES 4.9.

1. If $f(x) = ax^2 + bx + c$, show that $\Delta(f) = b^2 - 4ac$.
2. If $f(x) = x^3 + Ax + B$, show that $\Delta(f) = -4A^3 - 27B^2$.

COROLLARY 4.10. *$f(x)$ has a multiple root if and only if $\Delta(f) = 0$.*

THEOREM 4.11. *Suppose $f(x)$ and $g(x)$ are polynomials in $k[x]$. Then there are polynomials $F(x)$ and $G(x)$ in $k[x]$, such that*

$$R(f, g) = F(x)f(x) + G(x)g(x). \tag{4.4}$$

In particular, if $f(x)$, $g(x)$ are coprime, i.e., $(f(x), g(x)) = 1$, *then we can choose $F(x)$, $G(x)$ such that*

$$f(x)F(x) + g(x)G(x) = 1.$$

PROOF. Let

$$f(x) = a_0 + a_1 x + \cdots + a_n x^n \qquad (a_n \neq 0)$$

and

$$g(x) = b_0 + b_1 x + \cdots + b_m x^m \qquad (b_m \neq 0).$$

If $R(f, g) = 0$, there is nothing to prove. So let $R(f, g) = d \neq 0$.

Consider the system of equations

$$x^i f(x) = a_0 x^i + a_1 x^{i+1} + \cdots + a_n x^{i+n} \qquad (i = 0, 1, \ldots, m-1),$$

$$x^j g(x) = b_0 x^j + b_1 x^{j+1} + \cdots + b_m x^{j+m} \qquad (j = 0, 1, \ldots, n-1).$$

These equations can be rewritten as a single matrix equation $AX = Y$, where

$$A = \begin{pmatrix} a_0 & a_1 & \cdots & a_n & \\ & a_0 & \cdots & & a_n \\ & & \cdots & & \\ b_0 & b_1 & \cdots & b_m & \\ & b_0 & \cdots & & b_m \\ & & \cdots & & \end{pmatrix}, \qquad X = \begin{pmatrix} 1 \\ x \\ x^2 \\ \vdots \\ x^{m+n-1} \end{pmatrix}, \qquad Y = \begin{pmatrix} f(x) \\ f(x)x \\ \vdots \\ g(x) \\ g(x)x \\ \vdots \\ g(x)x^{n-1} \end{pmatrix}$$

The missing entries in A are all zeros. Clearly, $R(f, g) = \det(A) = d \neq 0$.

Since $d \neq 0$, $A^{-1} = (1/d)adjA$, where the matrix $adjA = (A_{ij})$ consists of the cofactors A_{ij} of A. Obviously, $X = (1/d)(adjA)Y$. Solving for the first coordinate of X, we obtain

$$d = \left(\sum_{j=1}^{m} A_{1j} x^{j-1}\right) f(x) + \left(\sum_{j=m+1}^{m+n} A_{1j} x^{j-m-1}\right) g(x).$$

Put

$$F(x) = \sum_{j=1}^{m} A_{1j} x^{j-1} \quad \text{and} \quad G(x) = \sum_{j=m+1}^{m+n} A_{1j} x^{j-m-1}.$$

To prove the last statement divide (4.4) throughout by $R(f, g) \neq 0$. □

COROLLARY 4.12. *Suppose $f(x)$ is a polynomial over k. Then there are polynomials $F(x)$, $G(x)$ in $k[x]$ such that the discriminant*

$$\Delta(f) = F(x)f(x) + G(x)f'(x). \tag{4.5}$$

PROOF. Since $\Delta(f) = (1/a_n)(-1)^{n(n-1)/2}R(f, f')$, we take $g(x) = f'(x)$ and replace $F(x)$, $G(x)$ by $cF(x)$ and $cG(x)$, respectively, where $c = (1/a_n)(-1)^{n(n-1)/2}$. □

4.4. Conjugate Fields

Throughout this section α stands for an algebraic number of degree n, i.e., the degree of the minimal polynomial $f(x)$ of α over $\mathbb{Q}$ is n. We can now prove the following results.

THEOREM 4.13. $\mathbb{Q}(\alpha) = \mathbb{Q}[\alpha]$.

PROOF. We have already seen that

$$\mathbb{Q}(\alpha) = \left\{ \frac{f_1(\alpha)}{f_2(\alpha)} \,\middle|\, f_j(x) \in \mathbb{Q}[x], \deg f_j(x) < n, j = 1, 2 \text{ and } f_2(\alpha) \neq 0 \right\}.$$

Thus it suffices to show that if $g(x)$ is a polynomial over $\mathbb{Q}$ of degree less than n such that $g(\alpha) \neq 0$, then $1/g(\alpha) = G(\alpha)$ for some $G(x)$ in $\mathbb{Q}[x]$. Since $f(x)$ is irreducible and deg $g(x) < n$, the g.c.d. $(f(x), g(x)) = 1$. Therefore, by Theorem 4.11,

$$f(x)F(x) + g(x)G(x) = 1,$$

for some $F(x)$, $G(x)$ in $\mathbb{Q}[x]$, so that $g(\alpha)G(\alpha) = 1$. □

Suppose $K = \mathbb{Q}(\alpha) = \mathbb{Q}[\alpha]$ is a number field of degree n and σ is an isomorphism of K into $\mathbb{C}$, i.e., σ: $K \to \mathbb{C}$ is an injective ring homomorphism. It is clear $\sigma(a) = a$ for all a in $\mathbb{Q}$. If

$$f(x) = a_0 + a_1 x + \cdots + a_{n-1}x^{n-1} + x^n$$

is the minimal polynomial of α over $\mathbb{Q}$, then

$$\sigma(f(\alpha)) = a_0 + a_1\sigma(\alpha) + \cdots + a_{n-1}(\sigma(\alpha))^{n-1} + (\sigma(\alpha))^n = 0,$$

showing that $\sigma(\alpha)$ is a conjugate of α. Thus σ permutes the conjugates of α. Since the conjugates $\alpha_1 = \alpha, \ldots, \alpha_n$ of α are all distinct, there are precisely n isomorphisms σ_i: $K \to \mathbb{C}$. These are uniquely determined by assigning a conjugate $\sigma_i(\alpha) = \alpha_i$ to α. Let $K^{(i)}$ be the image of K under σ_i, i.e., $K^{(i)} = \sigma_i(K)$. The fields $K^{(1)}, \ldots, K^{(n)}$ are called the *conjugates* of K. Note that $K^{(1)}, \ldots, K^{(n)}$ need not be distinct. Here are some examples.

EXAMPLES 4.14.

1. *Quadratic Fields.* Suppose $d \neq 0$, 1 is a square-free integer and $K = \mathbb{Q}(\sqrt{d}) = \mathbb{Q}[\sqrt{d}]$. The minimal polynomial of $\alpha = \sqrt{d}$ is $f(x) = x^2 - d$. It has two roots $\alpha = \alpha_1 = \sqrt{d}$, $\alpha_2 = -\sqrt{d}$. The two isomorphisms of K into $\mathbb{C}$ are $\sigma_1 = 1 = id$ and the *conjugation* $\alpha_2 = \sigma$ defined by $\sigma(x + y\sqrt{d}) = x - y\sqrt{d}$. Hence $K^{(1)} = K^{(2)} = K$.

2. *Cubic Fields.* Let $\alpha = \sqrt[3]{2}$ denote the real cube root of 2 and

$$\omega = \frac{-1 - \sqrt{-3}}{2},$$

a cube root of unity other than 1. The minimal polynomial of α is $x^3 - 2$ and the three conjugates of α are $\alpha = \omega^j \alpha$, $j = 0, 1, 2$. The field $K = \mathbb{Q}[\alpha]$ is contained in $\mathbb{R}$, whereas its conjugates $K^{(2)} = \mathbb{Q}[\omega\alpha]$ and $K^{(3)} = \mathbb{Q}[\omega^2\alpha]$ are not, so that $K \neq K^{(j)}$, for $j = 2, 3$.

3. *Cyclotomic Fields.* Suppose $p > 0$ is a prime and $\zeta = \zeta_p = e^{2\pi i/p} = \cos(2\pi/p) + i \sin(2\pi/p)$, i.e., ζ is a (primitive) pth root of unity. It can be easily checked (cf. Ref. 4) that the minimal polynomial of ζ over $\mathbb{Q}$ is

$$x^{p-1} + x^{p-2} + \cdots + x + 1.$$

Let $K = \mathbb{Q}(\zeta)$. The $p - 1$ conjugates of ζ are $\zeta, \zeta^2, \ldots, \zeta^{p-1}$ and since these are already in K, we must have $K^{(1)} = \cdots = K^{(p-1)}$. (See also Section 4.8.2.)

Definition 4.15. Suppose α is algebraic and $K = \mathbb{Q}(\alpha)$. The field extension $K/\mathbb{Q}$ is called *normal* or *galois* if

$$K^{(1)} = \cdots = K^{(n)}.$$

The quadratic and cyclotomic fields (Examples 4.14) provide us with some examples of galois extensions. The extension $\mathbb{Q}(\sqrt[3]{2})/\mathbb{Q}$ is not galois.

Suppose $K = \mathbb{Q}(\alpha)$ is a galois extension of $\mathbb{Q}$ and $\sigma: K \to \mathbb{C}$ is an isomorphism of K into $\mathbb{C}$. Then σ is an *automorphism of* K over $\mathbb{Q}$, i.e., an isomorphism of K onto itself which is the identity on $\mathbb{Q}$. The set

$$\mathrm{Gal}(K/\mathbb{Q}) = \{\sigma_1, \ldots, \sigma_n\}$$

of automorphisms of K over $\mathbb{Q}$ becomes a group under the composition of maps and is called the *galois group of K over* $\mathbb{Q}$. This group need not be abelian.

Definition 4.16. A galois extension $K = \mathbb{Q}(\alpha)$ of $\mathbb{Q}$ is *abelian* if its galois group $\mathrm{Gal}(K/\mathbb{Q})$ is abelian.

If $K = \mathbb{Q}(\sqrt{d})$ is a quadratic extension of $\mathbb{Q}$, then $\mathrm{Gal}(K/\mathbb{Q}) = \{1, \sigma\}$. Since any group of order two is abelian, $K/\mathbb{Q}$ is abelian. An example of a non-abelian extension of $\mathbb{Q}$ is $K = \mathbb{Q}(\alpha, \sqrt{-1})$, where α is the (real) fourth root of 2 (cf. Theorem 4.17 and Ref. 4, p. 200).

So far we have considered only the number fields $\mathbb{Q}(\alpha)$, the fields obtained by adjoining a single algebraic number α to $\mathbb{Q}$. Our next theorem shows that any number field is of this type.

Theorem 4.17. *Every algebraic number field K is a simple extension of $\mathbb{Q}$, i.e., $K = \mathbb{Q}(\gamma)$ for some algebraic number γ.*

PROOF. Since $K = \mathbb{Q}(\alpha_1, \ldots, \alpha_n) = \mathbb{Q}(\alpha_1, \ldots, \alpha_{n-1})(\alpha_n)$, the theorem will follow by induction on n if we can show that for a given pair of algebraic numbers α, β, we have $\mathbb{Q}(\alpha, \beta) = \mathbb{Q}(\gamma)$ for some γ.

Suppose $\{\alpha_1, \ldots, \alpha_m\}$ and $\{\beta_1, \ldots, \beta_n\}$ are the distinct roots of the minimal polynomials $f(x)$, $g(x)$ over $\mathbb{Q}$ of $\alpha = \alpha_1$, $\beta = \beta_1$, respectively. Since $\mathbb{Q}$ has infinitely many elements, there is a λ in $\mathbb{Q}$ such that

$$\lambda \neq \frac{\alpha_i - \alpha}{\beta - \beta_j} \qquad (i = 1, \ldots, m \quad \text{and} \quad j = 2, \ldots, n).$$

We put $\gamma = \alpha + \lambda\beta$. Clearly, $\mathbb{Q}(\gamma) \subseteq \mathbb{Q}(\alpha, \beta)$. Thus it suffices to show that $\mathbb{Q}(\alpha, \beta) \subseteq \mathbb{Q}(\gamma)$.

We may regard $g(x)$ and $h(x) = f(\gamma - \lambda x)$ as polynomials over $K = \mathbb{Q}(\gamma)$. Since $h(\beta) = f(\gamma - \lambda\beta) = f(\alpha) = 0$, the polynomial $x - \beta$ is a common factor in $\mathbb{C}[x]$ of $g(x)$, $h(x)$. In fact, $x - \beta$ is, up to a constant, the only common factor in $\mathbb{C}[x]$ of $g(x)$ and $h(x)$. For if there is another one, it has to be a multiple of $x - \beta_j$ for some $j = 2, \ldots, n$, so that $0 = h(\beta_j) = f(\gamma - \lambda\beta_j)$. Hence $\gamma - \lambda\beta_j = \alpha_i$ for some $i = 1, \ldots, m$. But $\gamma = \alpha + \lambda\beta$. Therefore,

$$\lambda = \frac{\alpha_i - \alpha}{\beta - \beta_j},$$

contradicting the choice of λ.

The minimal polynomial $\phi(x)$ over K of β has positive degree, divides both $g(x)$ and $h(x)$, and hence divides $x - \beta$. Therefore, $\phi(x) = x - \beta$, so that $\beta \in \mathbb{Q}(\gamma)$. Finally, $\alpha = \gamma - \lambda\beta$ is also in K. □

4.5. Algebraic Integers

We now associate to each number field K a ring $\mathcal{O}_K$ which is analogous to the ring $\mathbb{Z}$ for $\mathbb{Q}$. First we need the following definition:

DEFINITION 4.18. A nonempty set M is called a *module over a commutative ring A* with identity 1 or an *A-module* if

1. M is an Abelian group under (an operation we shall call) addition.
2. There is a scalar multiplication on M; i.e., given a *vector* $\alpha \in M$ and a *scalar* $a \in A$, there is a vector $a\alpha$ in M such that whenever $\alpha, \beta \in M$ and $a, b \in A$, we must have

   i. $(a + b)\alpha = a\alpha + b\alpha$,
   ii. $a(\alpha + \beta) = a\alpha + a\beta$,
   iii. $(ab)\alpha = a(b\alpha)$,
   iv. $1\alpha = \alpha$.

In particular, if A is a field then M is a *vector space* over A.

EXAMPLES 4.19.

1. An abelian group G is a $\mathbb{Z}$-module.
2. A commutative ring A with identity 1 is a module over itself.

A $\mathbb{Z}$-module M is a *finite* $\mathbb{Z}$-module or a *finitely generated* $\mathbb{Z}$-module if

$$\begin{aligned} M &= \mathbb{Z}\alpha_1 + \cdots + \mathbb{Z}\alpha_n \\ &= \{a_1\alpha_1 + \cdots + a_n\alpha_n \mid a_j \in \mathbb{Z}\} \end{aligned} \tag{4.6}$$

for some $\alpha_1, \ldots, \alpha_n$ in M. The set $\alpha_1, \ldots, \alpha_n$ is a *$\mathbb{Z}$-basis* for M if each α in M can be written in one and only one way as

$$\alpha = a_1\alpha_1 + \cdots + a_n\alpha_n \qquad (a_j \in \mathbb{Z}).$$

If M has a $\mathbb{Z}$-basis $\{\alpha_1, \ldots, \alpha_n\}$, then we say that M is a *free* $\mathbb{Z}$-module *of rank n*. In this case we write (4.6) as

$$M = \mathbb{Z}\alpha_1 \oplus \cdots \oplus \mathbb{Z}\alpha_n.$$

DEFINITION 4.20. An algebraic number α is *integral* or an *algebraic integer* if α is a root of a monic polynomial over $\mathbb{Z}$. If an algebraic integer α is in a number field K we say that α is an *integer of* (or *in*) K.

Note that if α is an algebraic integer, then so are its conjugates, for they all satisfy the same monic polynomial over $\mathbb{Z}$.

THEOREM 4.21. *Suppose $\alpha \in \mathbb{C}$. The following are equivalent*:

1. *α is integral*;
2. *$\mathbb{Z}[\alpha]$ is a finite $\mathbb{Z}$-module*;
3. *there is a finite $\mathbb{Z}$-module $M \neq \{0\}$ such that $\alpha M \subseteq M$.*

PROOF. The theorem is trivial if $\alpha = 0$. So we suppose that $\alpha \neq 0$. We shall show that $(1) \Rightarrow (2) \Rightarrow (3) \Rightarrow (1)$.

$(1) \Rightarrow (2)$. Since α is an algebraic integer,

$$\alpha^n = a_0 + a_1\alpha + \cdots + a_{n-1}\alpha^{n-1} \tag{4.7}$$

for some $a_0, a_1, \ldots, a_{n-1}$ in $\mathbb{Z}$. Consider the finite $\mathbb{Z}$-module

$$M = \mathbb{Z} + \mathbb{Z}\alpha + \cdots + \mathbb{Z}\alpha^{n-1}.$$

It is obvious that $M \subseteq \mathbb{Z}[\alpha]$ and it follows from (4.7) that $\alpha M \subseteq M$, which on repeated application shows that $\alpha^m \in M$ for all integers $m \geq 1$, i.e., $\mathbb{Z}[\alpha] \subseteq M$. Hence $\mathbb{Z}[\alpha] = M$.

$(2) \Rightarrow (3)$ is obvious with $M = \mathbb{Z}[\alpha]$.

$(3) \Rightarrow (1)$. Let

$$M = \mathbb{Z}\alpha_1 + \cdots + \mathbb{Z}\alpha_n \neq \{0\}$$

with $\alpha M \subseteq M$. Then for each $i = 1, \ldots, n$

$$\alpha\alpha_i = \sum_{j=1}^{n} a_{ij}\alpha_j \qquad (a_{ij} \in \mathbb{Z}). \tag{4.8}$$

We can rewrite (4.8) as a single matrix equation

$$\alpha\begin{pmatrix}\alpha_1\\ \vdots\\ \alpha_n\end{pmatrix}=\begin{pmatrix}a_{11} & \cdots & a_{1n}\\ \vdots & & \\ a_{n1} & \cdots & a_{nn}\end{pmatrix}\begin{pmatrix}\alpha_1\\ \vdots\\ \alpha_n\end{pmatrix},$$

or

$$(\alpha I - A)\mathbf{v} = \mathbf{0}, \tag{4.9}$$

where $A = (a_{ij}) \in M(n, \mathbb{Z})$, I is the $n \times n$ identity matrix, and

$$\mathbf{v} = \begin{pmatrix}\alpha_1\\ \vdots\\ \alpha_n\end{pmatrix}.$$

By our assumption, $M \neq \{0\}$, so that the vector $\mathbf{v} \neq \mathbf{0}$. Hence (4.9) is possible only if

$$\det(\alpha I - A) = 0. \tag{4.10}$$

The left-hand side of (4.10) is of the form

$$\alpha^n + a_{n-1}\alpha^{n-1} + \cdots + a_0$$

with $a_0, a_1, \ldots, a_{n-1}$ in $\mathbb{Z}$. This proves that α is an algebraic integer.

COROLLARY 4.22. *For an algebraic number field K, put*

$$\mathcal{O}_K = \{\alpha \in K \mid \alpha \text{ is an algebraic integer}\}.$$

Then $\mathcal{O}_K$ is a ring and $\mathcal{O}_K \supseteq \mathbb{Z}$.

PROOF. Suppose $\alpha, \beta \in \mathcal{O}_K$ not both zero. Then $\mathbb{Z}[\alpha]$, $\mathbb{Z}[\beta]$, and therefore $M = \mathbb{Z}[\alpha, \beta] \neq \{0\}$ are all finite $\mathbb{Z}$-modules. If γ is one of $\alpha - \beta$, $\alpha\beta$, then $\gamma M \subseteq M$, which proves that $\alpha - \beta$ and $\alpha\beta \in \mathcal{O}_K$. Therefore $\mathcal{O}_K$ is a ring. It is obvious that $\mathbb{Z} \subseteq \mathcal{O}_K$.

DEFINITION 4.23. The ring $\mathcal{O}_K$ is called the *ring of integers* of K.

EXERCISE 4.24. Show that $\mathcal{O}_{\mathbb{Q}} = \mathbb{Z}$.

THEOREM 4.25. *If $\alpha \in K$, then $a\alpha \in \mathcal{O}_K$ for a nonzero a in $\mathbb{Z}$.*

PROOF. If

$$\alpha^n + a_{n-1}\alpha^{n-1} + \cdots + a_1\alpha + a_0 = 0,$$

let a be the least common multiple of the denominators of $a_0, a_1, \ldots, a_{n-1}$. Then

$$(a\alpha)^n + aa_{n-1}(a\alpha)^{n-1} + \cdots + a^{n-1}a_1(a\alpha) + a^n a_0 = 0.$$

This shows that $a\alpha$ is a root of a monic polynomial over $\mathbb{Z}$. □

COROLLARY 4.26. *K is the quotient field of $\mathcal{O}_K$, i.e.,*

$$K = \left\{ \frac{\alpha}{\beta} \,\middle|\, \alpha, \beta \in \mathcal{O}_K \text{ with } \beta \neq 0 \right\}.$$

4.6. Integral Bases

Suppose $u_1, \ldots, u_n$ is an ordered basis of a number field K over $\mathbb{Q}$. In K, the multiplication by a fixed element α of K is a linear transformation $L = L_\alpha: K \to K$ over $\mathbb{Q}$. Therefore, L_α has a matrix $A_\alpha = A = (a_{ij})$ in $M(n, \mathbb{Q})$ with respect to the basis $\{u_1, \ldots, u_n\}$. It is defined by

$$\alpha u_i = \sum_{j=1}^{n} a_{ij} u_j.$$

If B is a matrix with respect to another ordered basis $\{v_1, \ldots, v_n\}$, then $B = P^{-1}AP$ for some P in $\mathrm{GL}(n, \mathbb{Q})$.

The *trace* $\mathrm{tr}(A)$ of an $n \times n$ square matrix $A = (a_{ij})$ is defined by

$$\mathrm{tr}(A) = a_{11} + \cdots + a_{nn}.$$

It is easy to check that $\mathrm{tr}(AB) = \mathrm{tr}(BA)$. Therefore, we have two well-defined functions from K to $\mathbb{Q}$:

1. The *trace*: $\mathrm{Tr}_{K/\mathbb{Q}}(\alpha) = \mathrm{tr}(A_\alpha)$

and

2. The *norm*: $N_{K/\mathbb{Q}}(\alpha) = \det(A_\alpha)$.

It is easily seen that (we drop the index $K/\mathbb{Q}$)

1. $\mathrm{Tr}(\alpha + \beta) = \mathrm{Tr}(\alpha) + \mathrm{Tr}(\beta)$;
2. $N(\alpha\beta) = N(\alpha)N(\beta)$;
3. If $\alpha \in \mathbb{Q}$, then $\mathrm{Tr}(\alpha) = n\alpha$ and $N(\alpha) = \alpha^n$.

Throughout the rest of this chapter $\sigma_1 = id, \ldots, \sigma_n$ will denote the n distinct $\mathbb{Q}$-isomorphisms of K into $\mathbb{C}$, i.e., the isomorphisms of K into $\mathbb{C}$ (which are identity on $\mathbb{Q}$).

THEOREM 4.27. *If $u_1, \ldots, u_n$ is an ordered basis of K over $\mathbb{Q}$, then the matrix $P = (\sigma_i(u_j))$ in $M(n, \mathbb{C})$ is nonsingular.*

PROOF. Let $K = \mathbb{Q}(\theta)$, with θ in $\mathcal{O}_K$. Then $1, \theta, \ldots, \theta^{n-1}$ is a basis of K over $\mathbb{Q}$. We first prove the theorem for this basis. The matrix P is nonsingular, because $\det(P)$ is the well-known *van der Monde determinant*

$$\begin{vmatrix} 1\sigma_1(\theta) & \cdots & \sigma_1(\theta)^{n-1} \\ 1\sigma_2(\theta) & \cdots & \sigma_2(\theta)^{n-1} \\ \vdots & & \\ 1\sigma_n(\theta) & \cdots & \sigma_n(\theta)^{n-1} \end{vmatrix} = \pm \prod_{\substack{i,j=1 \\ i<j}}^{n} [\sigma_i(\theta) - \sigma_j(\theta)],$$

which is nonzero in view of the fact that for $i \neq j$, $\sigma_i(\theta) \neq \sigma_j(\theta)$.

If $\{u_1, \ldots, u_n\}$ and $\{v_1, \ldots, v_n\}$ are any two bases of K over $\mathbb{Q}$, they are related by

$$v_j = \sum_{r=1}^{n} a_{rj} u_r,$$

where the matrix $A = (a_{ij})$ of transition is nonsingular. Hence $\det[\sigma_i(v_j)] = \det[\sigma_i(u_r)] \det(a_{rj})$, which proves that $\det[\sigma_i(v_j)] = 0$ if and only if $\det[\sigma_i(u_r)] = 0$. Hence by the first part of the proof, $\det[\sigma_i(u_j)] \neq 0$ for all bases $\{u_1, \ldots, u_n\}$. □

THEOREM 4.28. *If* $\alpha \in K$, *then*

1. $\mathrm{Tr}_{K/\mathbb{Q}}(\alpha) = \sigma_1(\alpha) + \cdots + \sigma_n(\alpha)$

and

2. $N_{K/\mathbb{Q}}(\alpha) = \sigma_1(\alpha) \ldots \sigma_n(\alpha)$.

PROOF. Apply σ_i to

$$\alpha u_j = \sum_{r=1}^{n} a_{rj} u_r \qquad (a_{rj} \in \mathbb{Q})$$

to get

$$\sigma_i(\alpha)\sigma_i(u_j) = \sum_{r=1}^{n} \sigma_i(u_r) a_{rj}. \tag{4.11}$$

If we define two $n \times n$ matrices by $D = \mathrm{diag}(\sigma_1(\alpha), \ldots, \sigma_n(\alpha))$ and $P = (\sigma_i(u_j))$, then (4.11) shows that $DP = PA_\alpha$, i.e., $D = PA_\alpha P^{-1}$. Hence

$$\begin{aligned} \mathrm{Tr}(\alpha) &= \mathrm{tr}(A_\alpha) = \mathrm{tr}(PA_\alpha P^{-1}) = \mathrm{tr}(D) \\ &= \sigma_1(\alpha) + \cdots + \sigma_n(\alpha) \end{aligned}$$

and similarly

$$\begin{aligned} N(\alpha) &= \det(A_\alpha) = \det(D) \\ &= \sigma_1(\alpha) \ldots \sigma_n(\alpha). \end{aligned}$$

□

COROLLARY 4.29. *If* $\alpha \in \mathcal{O}_K$, *then* $\mathrm{Tr}(\alpha)$ *and* $N(\alpha)$ *are in* $\mathbb{Z}$.

PROOF. Since each $\sigma_i(\alpha)$ is an algebraic integer, $N(\alpha)$ and $\mathrm{Tr}(\alpha)$ are in $\mathcal{O}_{\mathbb{Q}} = \mathbb{Z}$. □

THEOREM 4.30. *If* $K/\mathbb{Q}$ *is a finite extension of degree n, then* $\mathcal{O}_K$ *is a free* $\mathbb{Z}$*-module of rank n.*

PROOF. If $\alpha_1, \ldots, \alpha_n$ is a $\mathbb{Q}$-basis for K, then so is $a\alpha_1, \ldots, a\alpha_n$ for any nonzero a in $\mathbb{Z}$. Therefore, by Theorem 4.25, K has a $\mathbb{Q}$-basis consisting of algebraic integers $\alpha_1, \ldots, \alpha_n$. Then the entries of the $n \times n$ complex matrix $P = (\sigma_i(\alpha_j))$ are algebraic integers, so that the *discriminant* $\Delta = \Delta(\alpha_1, \ldots, \alpha_n)$ *of the basis* $\{\alpha_1, \ldots, \alpha_n\}$ defined by

$$\Delta = [\det(P)]^2$$

is an algebraic integer. But

$$\begin{aligned}\Delta &= [\det(P)]^2\\ &= \det(P'P)\\ &= \det[\mathrm{Tr}(\alpha_i\alpha_j)]\end{aligned}$$

is in $\mathbb{Q}$. Hence by Exercise 4.24 and Theorem 4.27, $\Delta \in \mathbb{Z}$ and $\Delta \neq 0$.

Among all the bases of $K/\mathbb{Q}$ consisting of algebraic integers we choose one, say $\omega_1, \ldots, \omega_n$ with the smallest $|\Delta(\omega_1, \ldots, \omega_n)|$. We shall show that $\omega_1, \ldots, \omega_n$ is a $\mathbb{Z}$-basis for $\mathcal{O}_K$. We can certainly write any α in $\mathcal{O}_K$ as

$$\alpha = a_1\omega_1 + \cdots + a_n\omega_n$$

with $a_1, \ldots, a_n$ in $\mathbb{Q}$. We show that all $a_j \in \mathbb{Z}$. Suppose not and without loss of generality let $a_1 \notin \mathbb{Z}$. We write $a_1 = a + r$ with a in $\mathbb{Z}$ and $0 < r < 1$. Define a new basis of $K/\mathbb{Q}$, consisting of algebraic integers, by

$$\begin{aligned}\alpha_1 &= \alpha - a\omega_1\\ &= (a_1 - a)\omega_1 + a_2\omega_2 + \cdots + a_n\omega_n,\\ \alpha_j &= \omega_j \qquad \text{for } j = 2, \ldots, n.\end{aligned}$$

The transition matrix between these two bases is

$$\left(\begin{array}{c|ccc} r & 0 & \cdots & 0 \\ \hline a_2 & & & \\ \vdots & & I_{n-1} & \\ a_n & & & \end{array}\right).$$

Hence $\Delta(\alpha_1, \ldots, \alpha_n) = r^2\Delta(\omega_1, \ldots, \omega_n)$, contradicting the minimality of $|\Delta(\omega_1, \ldots, \omega_n)|$. □

A $\mathbb{Z}$-basis of $\{\omega_1, \ldots, \omega_n\}$ of $\mathcal{O}_K$ is called an *integral basis* of K and its discriminant is the *discriminant* of K. The discriminant d_K of K is independent of $\{\omega_1, \ldots, \omega_n\}$, because any two such bases are related by a unimodular matrix. (Recall that the discriminant changes by the *square* of the determinant of the matrix.)

4.7. The Group of Units

We call $\mathcal{O}_K^\times$ the *group of units* (of the ring of integers) of K and now proceed to show that $\mathcal{O}_K^\times$ is finitely generated. First we make some preliminary remarks.

THEOREM 4.31. $\mathcal{O}_K^\times = \{u \in \mathcal{O}_K \mid N(u) = \pm 1\}$.

PROOF. If u is a unit, then $uv = 1$ for some v in $\mathcal{O}_K$, so that

$$N(u)N(v) = N(uv) = N(1) = 1.$$

Because $N(u), N(v) \in \mathbb{Z}$, we must have $N(u) = \pm 1$. Conversely, if

$$N(u) = u(\sigma_2(u) \cdots \sigma_n(u)) = \pm 1,$$

we take $v = \pm\sigma_2(u) \cdots \sigma_n(u)$ which is in $\mathcal{O}_K$. □

Let σ be one of $\sigma_1, \ldots, \sigma_n$. We call σ *real* or *complex* according as $\sigma(K)$ is contained in $\mathbb{R}$ or not. If for a complex σ we define its conjugate $\bar{\sigma}$: $K \to \mathbb{C}$ by $\bar{\sigma}(\alpha) = \overline{\sigma(\alpha)}$, then $\bar{\sigma} = \sigma_j \neq \sigma$ for some j. Thus the complex isomorphisms of $K \to \mathbb{C}$ occur in pairs. If σ is real (or, respectively, complex), then $\sigma(K)$ is called a *real* (respectively, *complex*) *conjugate* of K. If σ is complex, we call $\sigma(K)$, $\bar{\sigma}(K)$ a *pair of complex conjugate fields.*

Let r_1 be the number of real isomorphisms of K into $\mathbb{C}$ and r_2 the number of pairs of complex isomorphisms of K into $\mathbb{C}$. Then

$$n = r_1 + 2r_2.$$

THEOREM 4.32 (*Dirichlet*). *The group W_K of roots of unity in K is a finite cyclic group. If $r = r_1 + r_2 - 1$, then*

$$\mathcal{O}_K^\times \cong W_K \times \mathbb{Z}^r.$$

We first prove a couple of facts needed for the proof of this theorem.

The notion of the congruence modulo a nonzero integer m in $\mathbb{Z}$ can be generalized to any commutative ring A with identity. Suppose a is a nonzero element of A. For α, β in A, $\alpha \equiv \beta \pmod{a}$ if $\alpha - \beta = ab$ for some b in A. This is again an equivalence relation and partitions A into a disjoint union of inequivalent congruence classes. In particular, let

$$A = \mathcal{O}_K = \mathbb{Z}\omega_1 \oplus \cdots \oplus \mathbb{Z}\omega_n.$$

If $a \in \mathbb{Z} \subseteq \mathcal{O}_K$, $a > 0$, then any α in $\mathcal{O}_K$ is congruent $(\bmod\, a)$ to some

$$\beta = r_1\omega_1 + \cdots + r_n\omega_n \qquad (r_j \in \mathbb{Z}, 0 \le r_j < a).$$

Thus there are at most a^n inequivalent congruence classes.

Two elements α, β of A are called *associates* if $\alpha = \varepsilon\beta$ for a unit ε in A. We now prove the following theorem.

THEOREM 4.33. *For any positive integer a in $\mathbb{Z}$, there are only finitely many nonassociate α in $\mathcal{O}_K$ such that $|N(\alpha)| = a$.*

PROOF. We first note that for any nonzero α in $\mathcal{O}_K$, $N(\alpha)/\alpha \in \mathcal{O}_K$, for it is the product of all the conjugates α_j of α except $\alpha = \alpha_1$. Since each α_j is an algebraic integer, so is their product $N(\alpha)/\alpha$, which is clearly in K.

There are only finitely many congruence classes modulo a. So to prove the theorem, it suffices to show that if α, β are in the same congruence class and $|N(\alpha)| = |N(\beta)| = a > 0$, then α, β are associates. Clearly α, β are nonzero and since $\alpha - \beta = a\gamma$ for some γ in $\mathcal{O}_K$, $\alpha/\beta = 1 \pm [N(\beta)/\beta]\gamma \in \mathcal{O}_K$. Similarly $\beta/\alpha \in \mathcal{O}_K$. Hence α/β is a unit, i.e., α, β are associates. □

A subset of $\mathbb{R}^m$ is *discrete* if for each constant $C > 0$ there are only finitely many vectors $(x_1, \ldots, x_m)$ in S with

$$\max_{1 \le j \le m} |x_j| \le C.$$

THEOREM 4.34. *Any discrete subgroup Γ of $\mathbb{R}^m$ is a free $\mathbb{Z}$-module of rank $r \le m$, i.e.,*

$$\Gamma = \mathbb{Z}\gamma_1 \oplus \cdots \oplus \mathbb{Z}\gamma_r$$

with $r \le m$.

PROOF. The proof is by induction on m. If $m = 1$, $\mathbb{R}^m = \mathbb{R}$ and Γ being discrete, we can choose the smallest positive γ_1 in Γ (unless $\Gamma = \{0\}$ in which case there is nothing to prove). Then $\Gamma = \mathbb{Z}\gamma_1$. To prove this we have only to show that $\Gamma \subseteq \mathbb{Z}\gamma_1$. Write any γ in Γ as $\gamma = q\gamma_1 + r (q \in \mathbb{Z}, 0 \le r < \gamma_1)$. Then $r = \gamma - q\gamma_1 \in \Gamma$. The minimality condition on γ_1 implies that $r = 0$. Therefore, $\gamma = q\gamma_1 \in \mathbb{Z}\gamma_1$.

Suppose now that the theorem holds for all subgroups of $\mathbb{R}^{m-1}$ $(m > 1)$. If $\Gamma \neq \{\mathbf{0}\}$ is a discrete subgroup in $\mathbb{R}^m$, let Γ_1 be the (discrete) subgroup of $\mathbb{R}^{m-1} \subseteq \mathbb{R}^m$ generated by a nonzero element of Γ, i.e., $\Gamma_1 = \mathbb{Z}\gamma_1$. Since $\gamma_1 \neq \mathbf{0}$, we may relabel the axes and suppose that its first coordinate is nonzero. If $\mathbf{e}_1 = (1, 0, \ldots, 0), \ldots, \mathbf{e}_m = (0, \ldots, 0, 1)$ denotes the standard basis for $\mathbb{R}^m$, it is clear that $\{\gamma_1, \mathbf{e}_2, \ldots, \mathbf{e}_m\}$ is a basis for $\mathbb{R}^m$. Define an $\mathbb{R}$-linear transformation $L: \mathbb{R}^m \to \mathbb{R}^{m-1}$ by

$$L(x_1\gamma_1 + x_2\mathbf{e}_2 + \cdots + x_m\mathbf{e}_m) = (x_2, \ldots, x_m)$$

and put $\Gamma' = L(\Gamma)$. Then Γ' is a subgroup of $\mathbb{R}^{m-1}$. We claim that Γ' is discrete. To prove this let $C > 0$. Suppose $\gamma' = (x_2, \ldots, x_m) \in \Gamma'$ with

$$\max_{2 \leq j \leq m} |x_j| \leq C. \tag{4.12}$$

Then $\gamma' = L(\gamma)$ for some

$$\gamma = x_1\gamma_1 + x_2\mathbf{e}_2 + \cdots + x_m\mathbf{e}_m \qquad (0 \leq x_1 < 1) \tag{4.13}$$

in Γ. But if a denotes the largest coordinate (in absolute value) of γ_1, then the largest coordinate (in absolute value) of any γ in (4.13) is $\leq C + a$. Since Γ is discrete, there are only finitely many γ as in (4.13) for which $L(\gamma) = \gamma'$, i.e., (4.12) can hold for only finitely many γ' in Γ'. Hence Γ' is discrete.

By induction hypothesis, let $\gamma_2', \ldots, \gamma_r'$ $(r \leq m)$ be a $\mathbb{Z}$-basis for Γ'. Choose $\gamma_2, \ldots, \gamma_r$ in Γ with $L(\gamma_j) = \gamma_j'$ $(j = 2, \ldots, r)$. To prove the theorem, it suffices to show the following:

1. $\gamma_1, \ldots, \gamma_r$ are linearly independent;

and

2. $\Gamma = \mathbb{Z}\gamma_1 + \cdots + \mathbb{Z}\gamma_r$.

To prove (1) let $n_1\gamma_1 + \cdots + n_r\gamma_r = 0$. Then $L(n_1\gamma_1 + \cdots + n_r\gamma_r) = n_2\gamma_2' + \cdots + n_r\gamma_r' = 0$. But $\{\gamma_2', \ldots, \gamma_r'\}$ is a $\mathbb{Z}$-basis for Γ', so that $n_2 = \cdots = n_r = 0$. But then $n_1 = 0$ also. Hence $\gamma_1, \ldots, \gamma_r$ are linearly independent.

To prove (2), we first show that $\Gamma \subseteq \mathbb{Z}\gamma_1 + \cdots + \mathbb{Z}\gamma_r$. Let $\gamma \in \Gamma$. Since $\gamma' = L(\gamma) \in \Gamma'$, we have

$$\gamma' = n_2\gamma_2' + \cdots + n_r\gamma_r' \quad (n_j \in \mathbb{Z}, j = 2, \ldots, r),$$

so that

$$L(\gamma - (n_2\gamma_2 + \cdots + n_r\gamma_r)) = 0.$$

This shows that $\gamma - (n_2\gamma_2 + \cdots + n_r\gamma_r) \in \mathrm{Ker}(L) = \Gamma_1$, i.e., $\gamma - (n_2\gamma_2 + \cdots + n_r\gamma_r) = n_1\gamma_1$ for some n_1 in $\mathbb{Z}$. Hence $\Gamma \subseteq \mathbb{Z}\gamma_1 + \cdots + \mathbb{Z}\gamma_r$. The other inclusion $\mathbb{Z}\gamma_1 + \cdots + \mathbb{Z}\gamma_r \subseteq \Gamma$ is obvious. This proves (2). $\square$

The idea of the proof of Dirichlet's theorem is as follows. Suppose $K^{(1)}, \ldots, K^{(r_1)}$ are the real conjugates of K and $K^{(r_1+1)}$, $\bar{K}^{(r_1+1)} = K^{(r_1+r_2+1)}; \ldots; K^{(r_1+r_2)}, \bar{K}^{(r_1+r_2)} = K^{(r_1+2r_2)}$ are the r_2 pairs of complex conjugates of K, so that $n = r_1 + 2r_2$. For $\alpha \in K$ and $1 \leq i \leq n$ we denote by $\alpha^{(i)}$ the image of α in $K^{(i)}$ under the corresponding $\mathbb{Q}$-isomorphism of K into $\mathbb{C}$. Let $r = r_1 + r_2 - 1$. We define a group homomorphism $\lambda: \mathcal{O}_K^\times \to \mathbb{R}^r$ by

$$\lambda(\varepsilon) = (\log|\varepsilon^{(1)}|, \ldots, \log|\varepsilon^{(r)}|).$$

Then Dirichlet's theorem follows (by Theorem 2.34) from our next theorem.

THEOREM 4.35. (1) $\mathrm{Ker}(\lambda) = W_K$, *the group of roots of unity in* K, *that is a finite cyclic group*; (2) $\lambda(\mathcal{O}_K^\times)$ *is a lattice in* $\mathbb{R}^r$, i.e., *a free* $\mathbb{Z}$-*module of rank* r.

The proof of Theorem 4.35 we now give is from a course given at Göttingen by C. L. Siegel (cf. Refs. 5, 6). We need some technical lemmas.

LEMMA 4.36. *For any constant* $c > 0$, *there are only finitely many* α *in* $\mathcal{O}_K$ *such that*

$$|\sigma_i(\alpha)| \leq c, \qquad i = 1, \ldots, n. \tag{4.14}$$

PROOF. Let $\omega_1, \ldots, \omega_n$ be an integral basis for $\mathcal{O}_K$. Then for any

$$\alpha = x_1\omega_1 + \cdots + x_n\omega_n \qquad (x_j \in \mathbb{Z}) \tag{4.15}$$

in $\mathcal{O}_K$, we have

$$\sigma_i(\alpha) = x_1\sigma_i(\omega_1) + \cdots + x_n\sigma_i(\omega_n), \qquad i = 1, \ldots, n.$$

We rewrite these equations as a single matrix equation $\boldsymbol{\alpha} = P\mathbf{x}$, where

$$\boldsymbol{\alpha} = \begin{pmatrix} \sigma_1(\alpha) \\ \vdots \\ \sigma_n(\alpha) \end{pmatrix}, \qquad \mathbf{x} = \begin{pmatrix} x_1 \\ \vdots \\ x_n \end{pmatrix}$$

and $P = (\sigma_i(\omega_j))$ is the nonsingular matrix defined earlier, so that $\mathbf{x} = P^{-1}\boldsymbol{\alpha}$. Therefore, by assumption (4.14), for each j, $|x_j| \leq cc_1$ and the constant c_1 depends only on P and hence only on K. Consequently, there are only finitely many α as in (4.15) which can satisfy (4.14).

LEMMA 4.37. *Suppose* $A = (a_{ij})$ *is a real* $m \times n$ *matrix with* $m < n$. *Set*

$$a = \max_{1 \leq i \leq m} \sum_{j=1}^{n} |a_{ij}|.$$

If $t > 1$ is any integer, there exists a nonzero (column) vector $\mathbf{x}$ *in* $\mathbb{Z}^n$ *such that for* $\mathbf{y} = A\mathbf{x}$

$$\max_{1\le j\le n}|x_j| \le t \quad \textit{and} \quad \max_{1\le i\le m}|y_i| \le 2at^{1-n/m}, \tag{4.16}$$

where x_j, y_i *are the coordinates of* $\mathbf{x}$ *and* $\mathbf{y}$, *respectively.*

PROOF. Since $n > m$, for any $t > 1$,

$$t^{n/m} + 1 < (t+1)^{n/m},$$

so that we can choose an integer h such that

$$t^{n/m} \le h < (t+1)^{n/m}, \tag{4.17}$$

i.e.,

$$t^n \le h^m < (t+1)^n. \tag{4.18}$$

Subdivide each side of the cube

$$I = \{(y_1, \ldots, y_m) \in \mathbb{R}^m \,||y_i| \le at, \qquad i = 1, \ldots, m\}$$

into h equals parts, so that I has h^m subcubes each of side $2at/h$. There are $(t+1)^n$ points $\mathbf{x} = (x_1, \ldots, x_n)$ in $\mathbb{Z}^n$ with $x_j = 0, 1, \ldots, t$ and for each of these points $\mathbf{x}$, $|y_i| \le at$, so that $\mathbf{y} \in I$. Hence by (4.18), for two such (distinct) points $\mathbf{x}'$, $\mathbf{x}''$, the points $\mathbf{y}' = A\mathbf{x}'$, $\mathbf{y}'' = A\mathbf{x}''$ are in the same subcube of I. If $\mathbf{x} = \mathbf{x}' - \mathbf{x}''$, then the first inequality in (4.16) holds and by (4.17),

$$|y_i| \le 2at/h \le 2at^{1-n/m}. \qquad \square$$

Let $E = \{1, 2, \ldots, r_1 + r_2\} \subseteq \{1, 2, \ldots, n\}$. For r in E, let

$$\bar{r} = \begin{cases} r, & \text{if } 1 \le r \le r_1, \\ r + r_2, & \text{if } r_1 < r \le r_2. \end{cases}$$

For $X \subseteq E$, we put $\bar{X} = \{\bar{x} | x \in X\}$. Suppose E is a disjoint union of two nonempty subsets X, Y. We shall denote the number of elements in $X \cup \bar{X}$ by m. Clearly, $m < n$.

LEMMA 4.38. *There is a constant $c > 0$, depending only on K, such that given $t > 1$, there is a nonzero α in $\mathcal{O}_K$ with*

and
$$\left.\begin{aligned} c^{1-n}t^{1-n/m} \le |\alpha^{(k)}| \le ct^{1-n/m}, \quad & \textit{if } k \in X \\ c^{1-n}t \le |\alpha^{(l)}| \le ct, \quad & \textit{if } l \in Y. \end{aligned}\right\} \tag{4.19}$$

PROOF. Let $k_1, \ldots, k_u$ be the elements of X with $\bar{k}_i = k_i$, $l_1, \ldots, l_v$ those elements of X with $\bar{l}_i \ne l_i$. Then $m = u + 2v$. We now use Lemma 4.37 as follows: If

$$\mathcal{O}_K = \mathbb{Z}\omega_1 \oplus \cdots \oplus \mathbb{Z}\omega_n,$$

we define an $m \times n$ real matrix $A = (a_{ij})$ by

$$\begin{aligned} a_{i,j} &= \omega_j^{(k_i)}, \qquad 1 \le i \le u; \\ \left.\begin{aligned} a_{u+i,j} &= \operatorname{Re} \omega_j^{(l_i)}, \\ a_{u+v+i,j} &= \operatorname{Im} \omega_j^{(l_i)}. \end{aligned}\right\} & \qquad 1 \le i \le v. \end{aligned}$$

By Lemma 4.37, we choose a nonzero vector $\mathbf{x}$ in $\mathbb{Z}^n$ such that if $\mathbf{y} = A\mathbf{x}$, then (4.16) hold. We take

$$\alpha = \sum_{j=1}^{n} x_j \omega_j \neq 0$$

in $\mathcal{O}_K$. For k_i, l_i as above,

$$\begin{aligned} |\alpha^{(k_i)}| &= \left| \sum_{j=1}^{n} x_j \omega_j^{(k_i)} \right| = \left| \sum_{j=1}^{n} a_{ij} x_j \right| \\ &= |y_i| \le 2at^{1-n/m} \end{aligned}$$

and

$$\begin{aligned} |\alpha^{(l_i)}| &\le \left| \sum_{j=1}^{n} x_j \operatorname{Re} \omega_j^{(l_i)} \right| + \left| \sum_{j=1}^{n} x_j \operatorname{Im} \omega_j^{(l_i)} \right| \\ &= \left| \sum_{j=1}^{n} a_{u+i,j} x_j \right| + \left| \sum_{j=1}^{n} a_{u+v+i,j} x_j \right| \\ &\le 4at^{1-n/m}. \end{aligned}$$

By our choice, $|x_j| \le t$ for all j, so that for $l \in Y$, we have

$$|\alpha^{(l)}| = \left| \sum_{j=1}^{n} x_j \omega_j^{(l)} \right| \le c_1 t,$$

where

$$c_1 = \max_{l} \sum_{j=1}^{n} |\omega_j^{(l)}|.$$

If we put $c = \max(4a, c_1)$, we obtain the right sides of the inequalities in (4.19).

To prove the other halves of the inequalities, first note that $\{1, \ldots, n\}$ is a disjoint union of $X \cup \bar{X}$ and $Y \cup \bar{Y}$. Therefore α being an algebraic integer, we have

(1) For k in X,

$$\begin{aligned} 1 \le |N(\alpha)| &= \left(\prod_{i \in X \cup \bar{X}} |\alpha^{(i)}| \right) \left(\prod_{l \in Y \cup \bar{Y}} |\alpha^{(l)}| \right) \\ &\le |\alpha^{(k)}| (ct^{1-n/m})^{m-1} (ct)^{n-m} \\ &= |\alpha^{(k)}| c^{n-1} t^{n/m-1}, \end{aligned}$$

hence $|\alpha^{(k)}| \ge c^{1-n} t^{1-n/m}$;

(2) For l in Y,

$$1 \leq |N(\alpha)| = \left(\prod_{k \in X \cup \bar{X}} |\alpha^{(k)}|\right)\left(\prod_{j \in Y \cup \bar{Y}} |\alpha^{(j)}|\right)$$
$$\leq (ct^{1-n/m})^m |\alpha^{(l)}| (ct)^{n-m-1},$$

which gives $|\alpha^{(l)}| \geq c^{1-n} t$. □

LEMMA 4.39. *There is a sequence of nonzero algebraic integers α_ν in $\mathcal{O}_K$, $\nu = 1, 2, 3, \ldots$, such that for a constant c as in Lemma 4.38,*

$$|\alpha_\nu^{(k)}| > |\alpha_{\nu+1}^{(k)}|, \qquad \text{for } k \text{ in } X,$$
$$|\alpha_\nu^{(l)}| < |\alpha_{\nu+1}^{(l)}|, \qquad \text{for } l \text{ in } Y$$

and

$$|N(\alpha_\nu)| \leq c^n.$$

PROOF. Choose an integer M such that $M > c^n$ and $M^{n/m-1} > c^n$. Let $t_1 > 1$ and $t_{\nu+1} = Mt_\nu$. For each ν let α_ν be a nonzero algebraic integer for t_ν given by Lemma 4.38. Then

$$|\alpha_{\nu+1}^{(k)}| \leq ct_{\nu+1}^{1-n/m} = c(Mt_\nu)^{1-n/m}$$
$$< c^{1-n} t_\nu^{1-n/m} \leq |\alpha_\nu^{(k)}|$$

and

$$|\alpha_{\nu+1}^{(l)}| \geq c^{1-n} t_{\nu+1} = c^{1-n} M t_\nu$$
$$> ct_\nu \geq |\alpha_\nu^{(l)}|.$$

Finally,

$$|N_{K/\mathbb{Q}}(\alpha_\nu)| = \left(\prod_{k \in X \cup \bar{X}} |\alpha_\nu^{(k)}|\right)\left(\prod_{l \in Y \cup \bar{Y}} |\alpha_\nu^{(l)}|\right)$$
$$\leq (ct_\nu^{1-n/m})^m (ct_\nu)^{n-m} = c^n.$$

□

LEMMA 4.40. *There is a unit ε in $\mathcal{O}_K$ such that $|\varepsilon^{(k)}| < 1$ for k in X and $|\varepsilon^{(l)}| > 1$ for l in Y*

PROOF. In Lemma 4.39, since α_ν are nonzero algebraic integers, $|N(\alpha_\nu)|$ are positive integers bounded by c^n. By Theorem 4.33, there are only finitely many nonassociate α in $\mathcal{O}_K$ with $N(\alpha) \leq c^n$, hence $\alpha_\mu = \varepsilon\alpha_\nu\,(\mu > \nu)$ for some ε in $\mathcal{O}_K^\times$, so that

$$|\varepsilon^{(j)}| = \left|\frac{\alpha_\mu^{(j)}}{\alpha_\nu^{(j)}}\right|$$

is <1 if $j \in X$ and >1 if $j \in Y$.

LEMMA 4.41. *Suppose* $A = (a_{ij})$ *is a real* $r \times r$ *matrix such that*

1. $a_{ii} > 0$ *for each* $i = 1, \ldots, r$;
2. $a_{ij} \leq 0$ *for* $i \neq j$;
3. $\sum_{j=1}^{r} a_{ij} > 0$, *for each* $i = 1, \ldots, r$.

Then $det(A) \neq 0$.

PROOF. Suppose to the contrary that $\det(A) = 0$. Then for a nonzero column vector

$$\mathbf{t} = \begin{pmatrix} t_1 \\ \vdots \\ t_r \end{pmatrix},$$

we must have $A\mathbf{t} = \mathbf{0}$. Therefore, if

$$t_s = \max_{1 \leq j \leq r} |t_j|,$$

from

$$\sum_{j=1}^{r} a_{sj} t_j = 0,$$

we obtain

$$a_{ss} = |a_{ss}| = \left| \sum_{\substack{j=1 \\ j \neq s}}^{r} a_{sj} t_j / t_s \right|$$

$$\leq \sum_{j \neq s} |a_{sj}| = -\sum_{j \neq s} a_{sj},$$

so that

$$\sum_{j=1}^{r} a_{sj} \leq 0.$$

This contradiction to part 3 of the hypothesis proves the lemma. □

PROOF OF THEOREM 4.35.

(1) First we show that $\operatorname{Ker}(\lambda) = (\mathcal{O}_K^\times)_{\text{tor}}$. If $\varepsilon \in \operatorname{Ker}(\lambda)$, then $\log|\varepsilon^{(j)}| = 0$, i.e., $|\varepsilon^{(j)}| = 1$, for $j = 1, \ldots, r = r_1 + r_2 - 1$. But for $j = 1, \ldots, r_2$, we have $|\varepsilon^{(r_1+r_2+j)}| = |\varepsilon^{(r_1+j)}|$, so that $|\varepsilon^{(j)}| = 1$ with the exception of perhaps $j = r_1 + r_2, n$. Now

$$1 = |N(\varepsilon)| = \prod_{j=1}^{n} |\varepsilon^{(j)}| = |\varepsilon^{(r_1+r_2)}|^2 = |\varepsilon^{(n)}|^2$$

shows that $|\varepsilon^{(j)}| = 1$ for all $j = 1, \ldots, n$ and by Lemma 4.36, $\mathrm{Ker}(\lambda)$ is a finite group and therefore $\mathrm{Ker}(\lambda) \subseteq (\mathcal{O}_K^\times)_{\mathrm{tor}}$. Conversely, if $u \in (\mathcal{O}_K^\times)_{\mathrm{tor}}$, then $u^t = 1$ for some $t \geq 1$, so that for each $j = 1, \ldots, n$, $|u^{(j)}|^t = 1$, i.e., $|u^{(j)}| = 1$. Hence $u \in \mathrm{Ker}(\lambda)$.

Further, it is obvious that $(\mathcal{O}_K^\times)_{\mathrm{tor}} = W_K$. We must show that W_K is cyclic. Let

$$W_K = \{\exp(2\pi\sqrt{-1}a_j/b_j) | j = 1, \ldots, w\}.$$

If for $b = b_1 \ldots b_w$ and $\zeta = e^{2\pi\sqrt{-1}/b}$, we set

$$Z = \{n \in \mathbb{Z} | \zeta^n \in W_K\},$$

then Z is a subgroup of $\mathbb{Z}$. Since the only subgroups of $\mathbb{Z}$ are of the form $m\mathbb{Z}$, W_K is generated by ζ^m.

(2) First we show that $\lambda(\mathcal{O}_K^\times)$ is discrete. Given $c > 0$, we must show that there are only finitely many ε in $\mathcal{O}_K^\times$ such that

$$-c \leq \log|\varepsilon^{(j)}| \leq c,$$

i.e.,

$$e^{-c} \leq |\varepsilon^{(j)}| \leq e^c, \ j = 1, \ldots, r. \tag{4.20}$$

Let ε be any unit in $\mathcal{O}_K$ satisfying (4.20). Because $|N(\varepsilon)| = 1$, it follows that for all $j = 1, \ldots, n$, we have

$$|\varepsilon^{(j)}| \leq e^{nc}.$$

By Lemma 4.36, there are only finitely many such ε in $\mathcal{O}_K$. Therefore $\lambda(\mathcal{O}_K^\times)$ is discrete and by Theorem 4.34,

$$\lambda(\mathcal{O}_K^\times) = \mathbb{Z}\gamma_1 \oplus \cdots \oplus \mathbb{Z}\gamma_s, \ s \leq r.$$

It only remains to show that $s = r$. We shall show that $\lambda(\mathcal{O}_K^\times)$ contains a basis of $\mathbb{R}^r$ as a vector space over $\mathbb{R}$, which would not be possible if $s < r$.

For α in $K^\times$, we put

$$l^{(j)}(\alpha) = \begin{cases} \log|\alpha^{(j)}|, & j = 1, \ldots, r_1, \\ 2\log|\alpha^{(j)}|, & j = r_1 + 1, \ldots, r_1 + r_2. \end{cases} \tag{4.21}$$

Then

$$\log|N_{K/\mathbb{Q}}(\alpha)| = \sum_{j=1}^{r_1+r_2} l^{(j)}(\alpha).$$

Hence for any ε in $\mathcal{O}_K^\times$,

$$\sum_{i=1}^{r_1+r_2} l^{(j)}(\varepsilon) = 0.$$

Given $\varepsilon_1, \ldots, \varepsilon_r$ in $\mathcal{O}_K^\times$, consider the $r \times r$ real matrix $A = (l^{(j)}(\varepsilon_i))$. The vectors $\lambda(\varepsilon_1), \ldots, \lambda(\varepsilon_r)$ are linearly independent in $\mathbb{R}^r$ if and only if

$$\det(A) = 2^{r_2 - 1} \det(\log|\varepsilon_i^{(j)}|) \neq 0.$$

We apply Lemmas 4.40 and 4.41 as follows:

For each $1 \leq i \leq r$, let $Y = \{i\}$ and $X = E - Y$. By Lemma 4.40, choose ε_i in $\mathcal{O}_K^\times$ such that

$$|\varepsilon_i^{(i)}| > 1 \quad \text{and} \quad |\varepsilon_i^{(j)}| < 1, \text{ if } j \neq i.$$

Note that the $r \times r$ real matrix $A = (a_{ij}) = (l^{(j)}(\varepsilon_i))$ satisfies the hypothesis of Lemma 4.41. For clearly $a_{ii} > 0$, $a_{ij} < 0$ or $i \neq j$ and since

$$\sum_{j=1}^{r_1+r_2} l^{(j)}(\varepsilon_i) = 0,$$

we have for each $i = 1, \ldots, r$,

$$\sum_{j=1}^{r} a_{ij} = \sum_{j=1}^{r} l^{(j)}(\varepsilon_i) = -l^{(r_1+r_2)}(\varepsilon_i) > 0.$$

Therefore, by Lemma 4.41, $\det(A) \neq 0$ and $\lambda(\varepsilon_i), \ldots, \lambda(\varepsilon_r)$ are linearly independent. □

Let ζ be a generator of the finite cyclic group W_K of order w. We have established the existence of units $\varepsilon_1, \ldots, \varepsilon_r$ such that any ε in $\mathcal{O}_K^\times$ can be uniquely written as

$$\varepsilon = \zeta^a \varepsilon_1^{a_1} \ldots \varepsilon_r^{a_r} (a_j \in \mathbb{Z}, 0 \leq a < w).$$

These units $\varepsilon_1, \ldots, \varepsilon_r$ is called *fundamental units* of K.

For a generalization of Dirichlet's theorem, see Ref. 3, Theorem 3-3-11.

4.8. Quadratic and Cyclotomic Fields

4.8.1. Quadratic Fields

We now return to the diophantine equation.

$$x^2 - dy^2 = 1,$$

$d \neq 0, 1$ being a square-free integer. To solve this equation we apply Dirichlet's theorem to the quadratic field $K = \mathbb{Q}(\sqrt{d})$. We need to find an integral basis for $\mathcal{O}_K$.

THEOREM 4.42. *Let* $d \neq 0, 1$ *be a square-free integer and* $K = \mathbb{Q}(\sqrt{d})$. *Then* $1, \omega$ *is an integral basis of* $\mathcal{O}_K$ *with*

$$\omega = \begin{cases} \sqrt{d} & \text{if } d \equiv 2, 3 \pmod 4; \\ (1+\sqrt{d})/2 & \text{if } d \equiv 1 \pmod 4). \end{cases}$$

PROOF. For $d \equiv 1 \pmod 4$, $\omega = (1+\sqrt{d})/2$ is a root of the monic polynomial $f(X) = X^2 - \mathrm{Tr}(\omega)X + N(\omega) = X^2 - X - (d-1)/4$ in $\mathbb{Z}[X]$. Thus in any case, $\omega \in \mathcal{O}_K$, so that $\mathbb{Z} + \mathbb{Z}\omega \subseteq \mathcal{O}_K$.

Conversely, let $\alpha = x + y\sqrt{d}$ $(x, y \in \mathbb{Q})$ be an algebraic integer. Then $\mathrm{Tr}(\alpha) = 2x = m$ and $N(\alpha) = x^2 - dy^2$ are in $\mathbb{Z}$. If m is even, then $x = m/2$ and hence dy^2 are in $\mathbb{Z}$. But d is square-free, so y must also be in $\mathbb{Z}$ and $\alpha \in \mathbb{Z} + \mathbb{Z}\sqrt{d}$. If m is odd, then $dy^2 - 1/4 \in \mathbb{Z}$ and d being square-free, this is possible only if $y = n/2$ with n odd, which implies that $m^2 - dn^2 \equiv 0 \pmod 4$. Since m^2, n^2 are both $\equiv 1 \pmod 4$, we must have $d \equiv 1 \pmod 4$. Hence α is in

$$\mathbb{Z} + \mathbb{Z}\sqrt{d} \qquad \text{if } d \equiv 2, 3 \pmod 4;$$

$$\mathbb{Z}\frac{1}{2} + \mathbb{Z}\frac{\sqrt{d}}{2} = \mathbb{Z} + \mathbb{Z}\frac{1+\sqrt{d}}{2} \qquad \text{if } d \equiv 1 \pmod 4). \qquad \square$$

COROLLARY 4.43. *If* K *is a quadratic extension of* $\mathbb{Q}$ *and* d_K *is the discriminant of* K, *then* $K = \mathbb{Q}(\sqrt{d_K})$.

PROOF. In fact, if $K = \mathbb{Q}(\sqrt{d})$, then d_K is

$$[\det(\omega_j^{(i)})]^2 = \begin{vmatrix} 1 & \omega \\ 1 & \bar{\omega} \end{vmatrix}^2 = (\omega - \bar{\omega})^2$$

$$= \begin{cases} 4d & \text{if } d \equiv 2, 3 \pmod 4; \\ d, & \text{if } d \equiv 1 \pmod 4). \end{cases}$$

Hence in any case $K = \mathbb{Q}(\sqrt{d}) = \mathbb{Q}(\sqrt{d_K})$. $\square$

We now suppose that $d \equiv 2, 3 \pmod 4$, so that a nonzero element $\alpha = x + y\sqrt{d}$ of $\mathcal{O}_K = \mathbb{Z} + \mathbb{Z}\sqrt{d}$ is a unit if and only if

$$N(\alpha) = x^2 - dy^2 = \pm 1.$$

[On multiplying two units: $(x_1 + y_1\sqrt{d})(x_2 + y_2\sqrt{d}) = x_1x_2 + dy_1y_2 + (x_1y_2 + x_2y_1)\sqrt{d}$, it is obvious that the group G (of Exercise 2.4) is isomorphic to $A^{\times 2}$ or $A^{\times}$ according as $A^{\times}$ does or does not have an element of norm -1.] If $d > 0$, then $K = \mathbb{Q}(\sqrt{d})$ and its conjugate $\bar{K} = \mathbb{Q}(-\sqrt{d}) = K$ are both real fields, so that $r_1 = 2$, $r_2 = 0$ and $r = r_1 + r_2 - 1 = 1$. The only roots of unity in K are the real roots $1, -1$ of unity and $W_K = \{\pm 1\}$. By Dirichlet's theorem, choose η in $\mathcal{O}_K^{\times}$ such that for any u in $\mathcal{O}_K^{\times}$,

$$u = \pm \eta^n (n \in \mathbb{Z}).$$

It is clear that the generator η can be replaced by any of the four units: $\pm\eta$, $\pm\eta^{-1}$. Moreover, among these four generators, one and only one, say ε, is greater than 1. We call ε the *fundamental unit* of K.

In terms of solving (4.1) for $d > 1$, this amounts to putting $y = 1, 2, 3, \ldots$ in $1 + dy^2$ until for $y = y_1$ it becomes a square, say x_1^2 (with $x_1 > 0$). Note that this is guaranteed by Dirichlet's theorem. Then up to a change of sign of x, y, G is generated by (x_1, y_1).

The equation (4.1) is the simplest diophantine equation defined by a *norm form.* For further results on the norm form equations, see Ref. 7.

4.8.2. Cyclotomic Fields

Fields of the type $\mathbb{Q}(\zeta_m)$, where $\zeta_m = \cos(2\pi/m) + \sqrt{-1}\sin(2\pi/m)$ $(m \geq 1)$, a primitive mth root of unity, are called *cyclotomic fields.* (ζ is *primitive* if it is a generator of the group μ_m of mth roots of unity.) If r is the least common multiple of m, n, then $\mathbb{Q}(\zeta_r)$ contains both $\mathbb{Q}(\zeta_m)$ and $\mathbb{Q}(\zeta_n)$. Since all the conjugates of $\zeta = \zeta_m$ are powers of ζ, $K = \mathbb{Q}(\zeta)$ is a galois extension. For any σ in $\mathrm{Gal}(K/\mathbb{Q})$, $\sigma(\zeta)$ is a root of unity and hence $\sigma(\zeta) = \zeta^j$, for a unique $j = j(\sigma)$, $0 < j < m$. It is easy to check that the map

$$\Phi = \Phi_m\colon \mathrm{Gal}(K/\mathbb{Q}) \to (\mathbb{Z}/m\mathbb{Z})^\times$$

given by $\Phi(\sigma) = j(\sigma)$ is an injective group homomorphism. Hence, being isomorphic to a subgroup of $(\mathbb{Z}/m\mathbb{Z})^\times$, $\mathrm{Gal}(K/\mathbb{Q})$ is abelian. The map Φ is in fact onto, but this is much deeper.

As a closing remark, we now define a gaussian sum, a concept first introduced by Gauss, which plays an important role throughout number theory. Fix an odd prime p. The *gaussian sum* is

$$g = \sum_{x\in\mathbb{F}_p^\times} \left(\frac{x}{p}\right)\zeta^x, \tag{4.22}$$

where (x/p) is the Legendre symbol and $\zeta = \exp(2\pi\sqrt{-1}/p)$. Note that $g \in \mathbb{Q}(\zeta)$ and

$$g^2 = \sum_{x,\,y\in\mathbb{F}_p^\times} \left(\frac{xy}{p}\right)\zeta^{x+y}. \tag{4.23}$$

For a fixed x in $\mathbb{F}_p^\times$, the multiplication map $m_x(y) = xy$ is a permutation of $\mathbb{F}_p^\times$, so that in (4.23) we can replace the sum over y by the sum over xy to get

$$\begin{aligned} g^2 &= \sum_{x,y} \left(\frac{x^2y}{p}\right)\zeta^{x+xy} = \sum_{x,y}\left(\frac{y}{p}\right)\zeta^{x(1+y)} \\ &= \sum_{\substack{x,y\\ y\neq -1}} \left(\frac{y}{p}\right)\zeta^{x(1+y)} + \left(\frac{-1}{p}\right)(p-1). \end{aligned}$$

Because

$$1 + \zeta + \cdots + \zeta^{p-1} = 0,$$

for a fixed $y \neq -1$, the sum $\sum_x \zeta^{x(1+y)}$ is (up to a rearrangement of terms) equal to $\zeta + \cdots + \zeta^{p-1} = -1$. By Theorem 2.50,

$$g^2 = \left(\frac{-1}{p}\right)(p-1) - \sum_{y \neq -1} \left(\frac{y}{p}\right)$$

$$= \left(\frac{-1}{p}\right)p - \sum_{y \in \mathbb{F}_p^\times} \left(\frac{y}{p}\right) = \left(\frac{-1}{p}\right)p.$$

Thus we have proved the following theorem.

THEOREM 4.44. *The gaussian sum g defined by* (4.22) *satisfies*

$$g^2 = \left(\frac{-1}{p}\right)p.$$

COROLLARY 4.45. *Any quadratic extension* $K = \mathbb{Q}(\sqrt{d})$ *of* $\mathbb{Q}$ *is contained in* $\mathbb{Q}(\zeta)$ *for a root of unity* ζ.

PROOF. Put $\sqrt{-1} = i$. Since $2 = -i(1+i)^2$, $\sqrt{2} \in \mathbb{Q}(\zeta_8)$. For an odd prime p, $\sqrt{p} \in \mathbb{Q}(\zeta_p)$ or $\mathbb{Q}(\zeta_p, i)$ depending on whether $(-1/p) = 1$ or -1. So if $d = \pm 2^\alpha p_1 \dots p_r$ ($\alpha = 0, 1$ and p_j odd), it follows that

$$\sqrt{d} \in \mathbb{Q}(\zeta_8, \zeta_{p_1}, \dots, \zeta_{p_r}) \subseteq \mathbb{Q}(\zeta_m) \text{ for some } m. \qquad \square$$

Note that $\mathbb{Q}(\sqrt{d})$ is an abelian extension of $\mathbb{Q}$. Corollary 4.45 is a special case of a famous result conjectured by Kronecker and proved by Weber.

THEOREM 4.46 (*Kronecker–Weber*). *Any abelian extension of* $\mathbb{Q}$ *is contained in a cyclotomic extension* $\mathbb{Q}(\zeta)$.

For a proof, see Ref. 3, Chap. 13. When the *base field* $\mathbb{Q}$ is replaced by an *imaginary quadratic field* $K = \mathbb{Q}(\sqrt{d})$, $d < 0$, then the role of ζ is taken by the coordinates of points of finite order on certain elliptic curves; e.g., see Ref. 2, Chap. XIII. See also Ref. 8.

References

1. A. Baker, *Transcendental Number Theory*, Cambridge Univ. Press, Cambridge (1979).
2. J. W. S. Cassels and A. Fröhlich, *Algebraic Number Theory*, Academic, London (1967).
3. L. J. Goldstein, *Analytic Number Theory*, Prentice-Hall, Englewood Cliffs, New Jersey (1971).

4. S. Lang, *Algebra*, Addison-Wesley, Reading, Massachusetts (1970).
5. R. Narasimhan, S. Raghavan, S. S. Ranghachari, and Sunder Lal, *Algebraic Number Theory*, Mathematical Pamphlet No. 4, Tata Institute of Fundamental Research, Bombay (1966).
6. T. Ono, A course on Number Theory at the Johns Hopkins Univ., 1978 (unpublished).
7. W. M. Schmidt, Norm form equations, *Ann. Math.* **96**, 526–551 (1972).
8. G. Shimura, *Automorphic Functions and Number Theory*, Lecture Notes in Math. No. 54, Springer-Verlag, Berlin (1968).
9. A Weil, *Number Theory—An Approach through History*, Birkhäuser, Boston (1984).

5

Algebraic Curves

5.1. Introduction

So far we have considered only equations of degree at most 2. Because of the group structure on the integer solutions of $x^2 - dy^2 = 1$, we were able to employ algebraic methods to find these solutions. Let us now take, as an example, the diophantine equation

$$y(y-1) = x(x-1)(x+1). \tag{5.1}$$

To solve it in integers is to look for those products of two consecutive integers that are also products of three consecutive integers. The integers solutions of such cubics have, in general, no group structure. However, there is a very elegant group structure on the rational solutions of these equations. Moreover, as we shall see in the present chapter, many equations

$$f(x, y) = 0$$

of higher degree can be reduced to the cubic

$$y^2 = x^3 + Ax + B.$$

An equation $f(x, y) = 0$ represents a curve in the plane. It is, therefore, not surprising that the geometry of a curve is closely related to its arithmetic.

In general, for each j in an index set I, let $f_j(\mathbf{x}) \in \mathbb{Q}[x_1, \ldots, x_n]$. In algebraic geometry one studies the set X of common solutions in $\mathbb{C}^n$ of (possibly infinitely many) equations

$$f_j(\mathbf{x}) = 0, \qquad j \in I.$$

We call X a *variety defined over* $\mathbb{Q}$. If X is not a union of two proper varieties, X is called an *irreducible variety*. By Hilbert's basis theorem, any variety is defined, necessarily, by a finite set of equations. For the sake of further simplicity, we shall restrict ourselves to the variety defined by a single equation $f(x, y) = 0$.

5.2. Preliminaries

Suppose K/k is any field extension and $f(\mathbf{x}) \in k[x_1, \ldots, x_n]$. We say that $f(\mathbf{x})$ is *irreducible over K*, if $f(\mathbf{x})$ is not a product of two polynomials in $K[x_1, \ldots, x_n]$, each of degree at least one. Unless stated otherwise, irreducible will mean irreducible over k. The ring $k[x_1, \ldots, x_n]$ is a *unique factorization domain.* This means that any polynomial $f(\mathbf{x})$ in $k[x_1, \ldots, x_n]$ is a product

$$f(\mathbf{x}) = \prod_{j=1}^{r} p_j(\mathbf{x})^{m_j}, \qquad m_j \in \mathbb{N}, \tag{5.2}$$

of distinct irreducible polynomials $p_j(\mathbf{x})$ in $k[x_1, \ldots, x_n]$. The irreducible factors $p_j(\mathbf{x})$ are unique, except for a rearrangement of order and multiplication by nonzero constants.

In the representation (5.2), it is sometimes convenient to allow $m_j = 0$. For example, the *greatest common divisor* (f, g) of two nonzero polynomials

$$f(\mathbf{x}) = \prod_{j=1}^{r} p_j(\mathbf{x})^{m_j}, \qquad m_j \geq 0$$

and

$$g(\mathbf{x}) = \prod_{j=1}^{r} p_j(\mathbf{x})^{n_j}, \qquad n_j \geq 0$$

(in $k[x_1, \ldots, x_n]$) is by definition any polynomial that is a nonzero constant multiple of

$$\prod_{j=1}^{r} p_j(\mathbf{x})^{\min(m_j, n_j)}.$$

Unless stated otherwise, we shall have $k = \mathbb{Q}$, where $K \subseteq \mathbb{C}$. In particular, K will be either a number field (including $K = \mathbb{Q}$) or $K = \mathbb{R}, \mathbb{C}$. When a polynomial in $\mathbb{Q}[x_1, \ldots, x_n]$ is irreducible over $\mathbb{C}$, it is referred to as an *absolutely irreducible* polynomial.

5.3. Homogeneous Polynomials and Projective Spaces

Suppose $f(x_1, \ldots, x_n)$ in $\mathbb{Q}[x_1, \ldots, x_n]$ is a homogeneous polynomial of degree m, i.e.,

$$f(tx_1, \ldots, tx_n) = t^m f(x_1, \ldots, x_n).$$

A point $(a_1, \ldots, a_n)$ of $\mathbb{C}^n$ satisfies

$$f(x_1, \ldots, x_n) = 0 \tag{5.3}$$

if and only if $(ta_1, \ldots, ta_n)$ does, for all t in $\mathbb{C}$. Of course, $\mathbf{0} = (0, \ldots, 0)$ always satisfies (5.3) and is called its *trivial solution.* This leads to the concept of a projective space.

For any field K, let

$$K^n = \{(x_1, \ldots, x_n) \mid x_j \in K, j = 1, \ldots, n\}$$

be the vector space of n-tuples of elements of K. On the set

$$K^{n+1} - \{\mathbf{0}\} = \{\mathbf{x} \in K^{n+1} \mid \mathbf{x} \neq \mathbf{0}\}$$

define an equivalence relation by $\mathbf{x} \sim \mathbf{y}$ if and only if $\mathbf{y} = t\mathbf{x}$ for some t in $K^\times$. Then the set

$$\mathbb{P}^n(K) = K^{n+1} - \{\mathbf{0}\}/\sim$$

of equivalence classes is called the *projective space over* K, i.e., $\mathbb{P}^n(K)$ consists of all nonzero vectors in K^{n+1} with two vectors $\mathbf{x}$, $\mathbf{y}$ not considered different if $\mathbf{y} = t\mathbf{x}$ with t in $K^\times$. We can also think of $\mathbb{P}^n(K)$ as consisting of lines through the origin $\mathbf{0}$ of K^{n+1}. The *affine space* $\mathbb{A}^n(K) = K^n$ may be regarded as a subset of $\mathbb{P}^n(K)$ by the inclusion map

$$\mathbb{A}^n(K) \ni (x_1, \ldots, x_n) \to (x_1, \ldots, x_n, 1) \in \mathbb{P}^n(K).$$

5.4. Plane Algebraic Curves

A *curve* (or to be precise, a *plane algebraic curve*) C *defined over* $\mathbb{Q}$ is the set of solutions in $\mathbb{C}^2$ of a polynomial equation

$$f(x, y) = 0 \tag{5.4}$$

with $f(x, y) \in \mathbb{Q}[x, y]$. Suppose that

$$f(x, y) = \prod_{j=1}^{r} p_j(x, y)^{m_j}$$

is the factorization of $f(x, y)$ into irreducible factors. Then the curves C_j defined by

$$p_j(x, y) = 0, \qquad j = 1, \ldots, r$$

are all irreducible. The curve C is the union of $C_1, \ldots, C_r$, called the (*irreducible*) *components* of C. (Note that irreducible means irreducible over $\mathbb{Q}$.)

A *projective curve over* $\mathbb{Q}$ is the set of solutions in $\mathbb{P}^2(\mathbb{C})$ of a homogeneous equation

$$F(X, Y, Z) = 0 \tag{5.5}$$

with $F(X, Y, Z)$ in $\mathbb{Q}[X, Y, Z]$. One can *homogenize* (5.4) to get an equation like (5.5) by putting $x = X/Z$, $y = Y/Z$ and then multiplying throughout by $Z^{\deg(f)}$. Suppose now that this has been done, i.e., (5.5) has been obtained from (5.4) in this manner. We say that (5.5) is a *projective model* for the *affine curve* C defined by (5.4). The solutions of (5.4) are precisely those solutions of (5.5) for which $Z \neq 0$, because (a, b) is a point on (5.4) if and only if $(a, b, 1)$ is a point on (5.5). A point (a, b, c) on (5.5) with $c = 0$ is called a *point at infinity* on C. The *line at infinity* is the set of all points (X, Y, Z) in $\mathbb{P}^2(\mathbb{C})$ for which $Z = 0$. The points on (5.4) are called the *affine part* of the projective curve (5.5). A projective curve is *complete* in the sense that it has the points at infinity that are missing from its affine part. For this reason, we must work sometimes with the projective model only.

5.5. Singularities of a Curve

Let C be a projective curve defined by (5.5). We say that C is of *order* n if $\deg(F) = n$. We call C *linear, quadratic, cubic, quartic,* or *quintic* according as $n = 1, 2, 3, 4$, or 5. A point P on (5.5) is a *multiple point* of *multiplicity* $r \geq 1$ or an *r-fold* point if all the partial derivatives of $F(X, Y, Z)$ of order $<r$ vanish at P, but there is a partial derivative of order r that does not vanish at P. If $r = 1$, P is called a *regular* or a *nonsingular* point. (For a point P, not at infinity, to be a regular point is equivalent to at least one of dy/dx, dx/dy being well defined—i.e., to the curve having a unique tangent at P.) If $r > 1$, P is called a *singular* point or a *singularity* of C. We say that P is a *double* or a *triple* point according as $r = 2$ or 3. A curve is *singular* if it has a singularity, otherwise it is *nonsingular* or *smooth*.

EXAMPLES 5.1.

1. Let C be the affine curve defined by $y^2 = x^3$ (cf. Fig. 5.1). To homogenize this equation, we put $x = X/Z$, $y = Y/Z$ and obtain

$$F(X, Y, Z) = Y^2Z - X^3 = 0.$$

The singularities of C must satisfy

$$\frac{\partial F}{\partial X} = -3X^2 = 0,$$

$$\frac{\partial F}{\partial Y} = 2YZ = 0,$$

$$\frac{\partial F}{\partial Z} = Y^2 = 0,$$

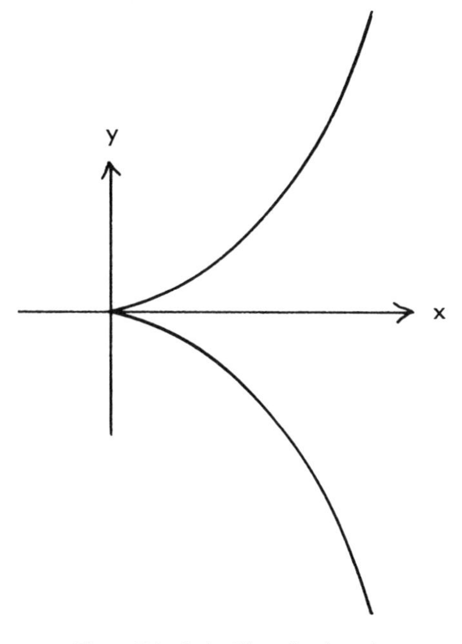

Figure 5.1. A double point (cusp).

which are therefore of the form $(0, 0, Z)$ with $Z \neq 0$. Thus $(0, 0, 1)$ is the only singularity of C. In fact, it is a double point, because

$$\left.\frac{\partial^2 F}{\partial Y^2}\right|_{(0,0,1)} \neq 0.$$

2. To locate the singularities of

$$F(X, Y, Z) = Y^2 Z - X^3 - X^2 Z = 0,$$

let us first consider its affine part (cf. Fig. 5.2)

$$y^2 = x^3 + x^2.$$

If $f(x, y) = y^2 - x^3 - x^2$, a point (x_0, y_0) is a singularity if and only if it satisfies

$$\frac{\partial f}{\partial x} = x(3x + 2), \qquad \frac{\partial f}{\partial y} = 2y = 0.$$

The origin $\mathbf{0} = (0, 0)$ is the only such point. Moreover, it is a double point. The only other singularity possible is at a point at infinity, i.e., if $Z = 0$, hence $X = 0$ and we may take $Y = 1$. But

$$\left.\frac{\partial F}{\partial Z}\right|_{(0,1,0)} \neq 0,$$

so that the point at infinity is not a singularity.

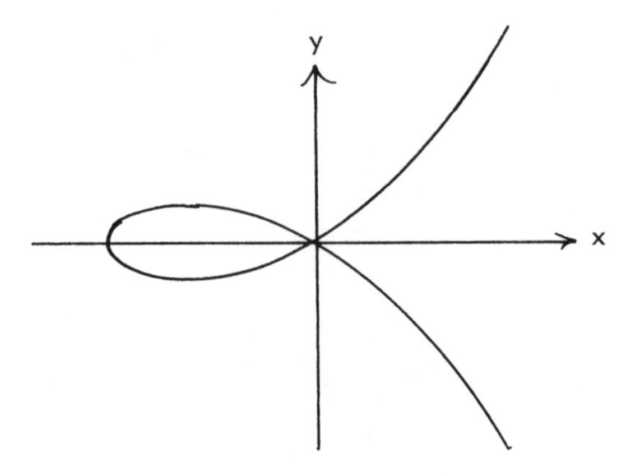

Figure 5.2. A double point (node).

EXERCISES 5.2.

1. Show that straight lines and irreducible conics have no singularity in $\mathbb{P}^2(\mathbb{C})$.

2. Show that the curve E defined by $y^2 = x^3 - x$ has no singularity (including at infinity).

3. Show that the origin $\mathbf{0} = (0, 0)$ is a triple point of the curve (cf. Fig. 5.3) defined by

$$(x^2 + y^2)^2 + 3x^2y - y^3 = 0.$$

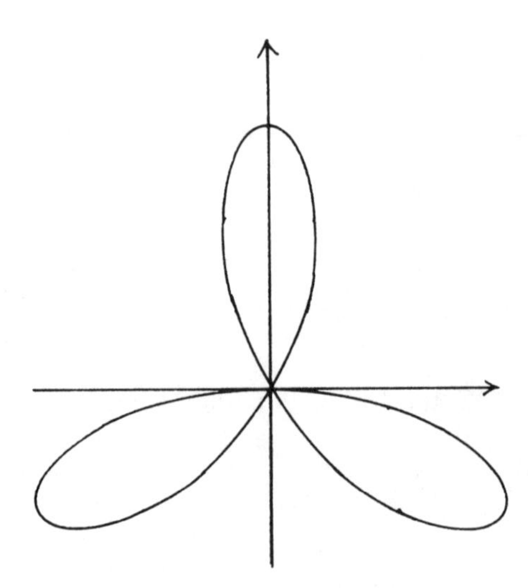

Figure 5.3. A triple point.

REMARKS 5.3.

1. An irreducible curve has only finitely many singularities (cf. Theorem 5.8).

2. A point common to two or more components of C is always a singularity of C. (The proof is left as an exercise.)

5.6. Birational Geometry

Suppose K is a subfield of $\mathbb{C}$ and C is an irreducible curve given by a polynomial equation

$$f(x, y) = 0$$

with coefficients in K. If L is another field with $K \subseteq L \subseteq \mathbb{C}$, a point $P = (x, y)$ on C is called an *L-rational point* (or simply a *rational point* in case $L = \mathbb{Q}$) if $x, y \in L$.

Our goal is to study the rational points on curves defined over $\mathbb{Q}$, which we may assume to be irreducible over $\mathbb{Q}$. The problem gets more complicated as the degree (or rather the genus, a term to be explained later,) of the curve increases. We start with the simplest curves. Unless stated otherwise, we shall assume that $K = L = \mathbb{Q}$. Consider the field of rational functions

$$\mathbb{Q}(x_1, \ldots, x_n) = \left\{ \phi(x_1, \ldots, x_n) = \frac{f(x_1, \ldots, x_n)}{g(x_1, \ldots, x_n)} \,\middle|\, f, g \in \mathbb{Z}[x_1, \ldots, x_n], g \neq 0 \right\}.$$

It is the quotient field of the ring of polynomials $\mathbb{Q}[x_1, \ldots, x_n]$. If $\phi = f/g \in \mathbb{Q}(x_1, \ldots, x_n)$ and $P = (a_1, \ldots, a_n)$ is a rational point; i.e., if all the $a_i \in \mathbb{Q}$, then $\phi(P) \in \mathbb{Q}$ [provided $g(P) \neq 0$].

DEFINITION 5.4. An irreducible curve C defined by $f(x, y) = 0$ is called a *rational curve* if there are rational functions $\phi_1(t)$, $\phi_2(t)$ in $\mathbb{Q}(t)$, such that

1. for almost all (i.e., for all except finitely many) $t \in \mathbb{C}$, $f(\phi_1(t), \phi_2(t)) = 0$; and
2. for almost all P on C, $P = (\phi_1(t), \phi_2(t))$, for some t in $\mathbb{C}$.

We say that C is *parametrized* by t.

EXAMPLES 5.5. In these examples the curve is parametrized by the slope t of the family of lines

$$y = tx.$$

(1) The unit circle (Fig. 5.4) $x^2 + (y-1)^2 = 1$ has a parametrization

$$x = \phi_1(t) = \frac{2t}{1+t^2},$$

$$y = \phi_2(t) = \frac{2t^2}{1+t^2}.$$

Note that $\phi_j(\sqrt{-1})$ is not defined. On the other hand, the point $(0, 2)$ on C is not $(\phi_1(t), \phi_2(t))$ for any t in $\mathbb{C}$.

(2) The singular cubic (Fig. 5.5) $y^2 = x^3$ has parametrization

$$x = \phi_1(t) = t^2,$$

$$y = \phi_2(t) = t^3.$$

(3) Another singular cubic (Fig. 5.6) $y^2 = x^2(x+1)$ may be parametrized as

$$x = \phi_1(t) = t^2 - 1,$$

$$y = \phi_2(t) = t(t^2 - 1).$$

Note that almost all rational points on these curves are obtained by taking t from $\mathbb{Q}$.

Let C_1 and C_2 be two curves such that the coordinates of almost all points on C_2 are rational functions of the coordinates of points on C_1 and vice versa. This means that there are rational functions

$$\phi_j(x, y) = \frac{f_j(x, y)}{g_j(x, y)} \in \mathbb{Q}(x, y), \qquad j = 1, 2$$

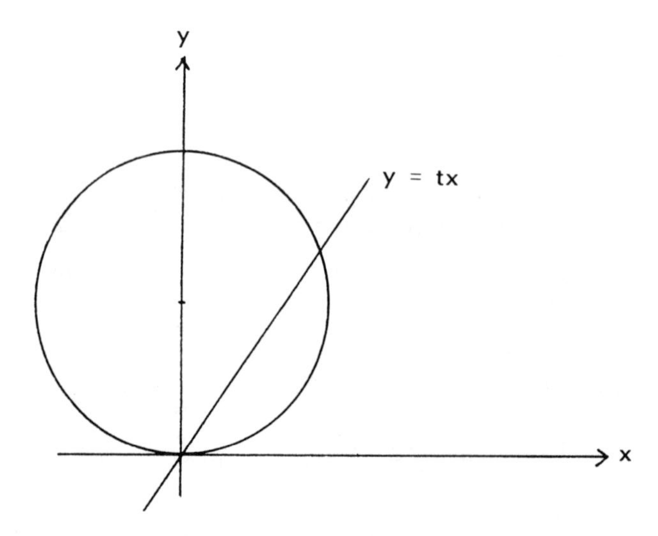

Figure 5.4. Parametrizing a circle.

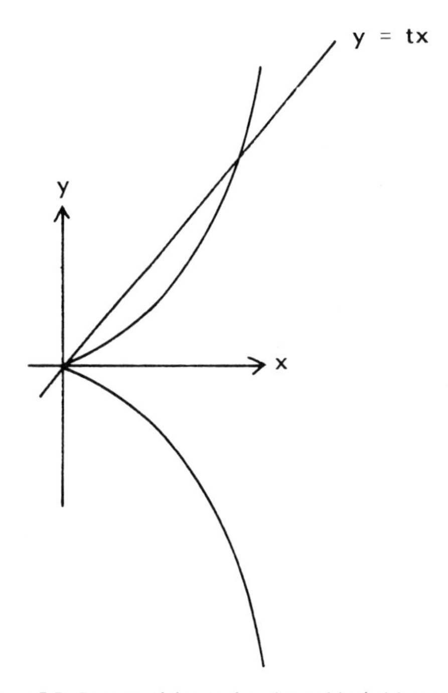

Figure 5.5. Parametrizing a singular cubic (with a cusp).

such that

1. for almost all P on C_1, $g_1(P)g_2(P) \neq 0$; and
2. the set $\{\Phi(P) = (\phi_1(P), \phi_2(P)) | P \in C_1, g_1(P)g_2(P) \neq 0\}$

accounts for almost all points on C_2.

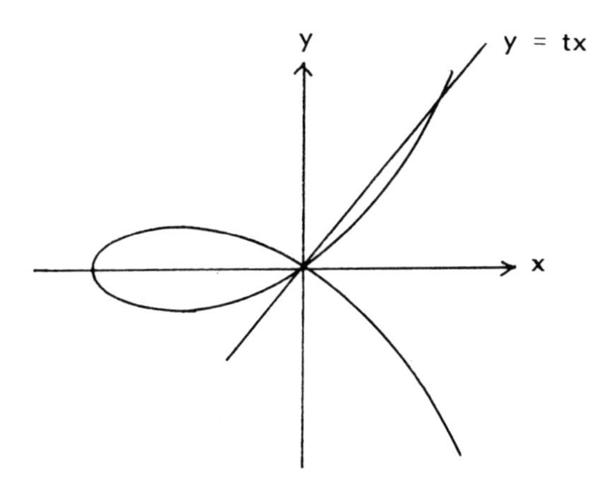

Figure 5.6. Parametrizing a singular cubic (with a node).

A similar statement holds in the other direction. Two such curves C_1 and C_2 are called *birationally equivalent* (*over* $\mathbb{Q}$ to be precise). If C_1 and C_2 are birationally equivalent, then almost all points on one curve are obtained from those on the other by the *rational maps*:

$$\begin{aligned} \Phi&: C_1 \to C_2 \\ \Psi&: C_2 \to C_1 \end{aligned} \tag{5.6}$$

given by

$$\Phi(P) = (\phi_1(P), \phi_2(P))$$

and

$$\Psi(Q) = (\psi_1(Q), \psi_2(Q)).$$

(5.6) is called a *birational correspondence* (*over* $\mathbb{Q}$) between C_1 and C_2 if $\Psi \circ \Phi: C_1 \to C_1$ and $\Phi \circ \Psi: C_2 \to C_2$ are the identity functions, whenever defined. Note that Φ and Ψ are defined for almost all points on C_1 and C_2, respectively.

EXAMPLE 5.6 (*Tate* [8]). Let C_1 be the Fermat curve given by

$$x^3 + y^3 = 1. \tag{5.7}$$

If in (5.7) we make the rational substitution

$$\begin{aligned} &x = \frac{6}{X} + \frac{Y}{6X}, \qquad y = \frac{6}{X} - \frac{Y}{6X}, \\ \Leftrightarrow \\ &X = \frac{12}{x+y}, \qquad Y = 36\frac{x-y}{x+y}, \end{aligned} \tag{5.8}$$

we get the curve C_2 with equation

$$Y^2 = X^3 - 432, \tag{5.9}$$

and (5.8) give a birational correspondence between C_1 and C_2.

REMARK 5.7. Birational equivalence between curves is an equivalence relation. If C_1 and C_2 are birationally equivalent, then C_1 has infinitely many rational points on it if and only if C_2 does. Thus, among a class of birationally equivalent curves, we may choose one that has the simplest form. This is what we shall do now.

5.7 Some Results from Algebraic Geometry

We need to recall some facts from algebraic geometry. Some of these results are quite deep, while others follow easily from them. One of the most useful is the following result on the intersection of two curves. We shall not prove the theorems marked with an asterisk. The proofs (for this chapter) can be found in Ref. 7. We must assume that our curves are projective and that $K = \mathbb{C}$, so that there are no missing points.

THEOREM 5.8* (*Bezout*). *Two projective curves of order m and n having no component in common intersect in mn points* (*counted properly*).

Thus two straight lines meet in one point (which may be the point at infinity). Note that this is false in the affine case. A straight line intersects a conic in two points and a cubic in three points. Two cubics intersect in nine points, etc. A point of intersection which is nonsingular for both the curves is counted once. A double point of either curve is counted at least twice. The number of times a point of intersection is to be counted is called its *intersection multiplicity*. For details see Refs. 2 or 4.

A projective curve of order n is given by a polynomial equation

$$F(X, Y, Z) = \sum_{i+j \leq n, i,j=0}^{n} a_{ij} X^i Y^j Z^{n-i-j} = 0, \qquad (5.10)$$

so that $F(X, Y, Z)$ has $(n+1)(n+2)/2$ terms. Some, but not all, of these may be zero. Two such polynomials F and G define the same curve if and only if $F = c \cdot G$ with a nonzero constant c. Thus the curves of order n may be identified with the points of the projective space $\mathbb{P}^m(\mathbb{C})$, where

$$m = \frac{(n+1)(n+2)}{2} - 1 = \frac{n(n+3)}{2}.$$

There is always a curve of order n passing through any given set of $n(n+3)/2$ points, because the coefficients of the polynomial in (5.10) can be taken as a nonzero solution of m linear equations [obtained by substituting these m points into (5.10)] in the $m+1$ arbitrary coefficients. If the coefficients of the highest terms are all zero, we can multiply by a suitable power of, say, X to get an equation of the desired degree. For later use, we record it as the following theorem.

THEOREM 5.9. *There is always a curve of degree n passing through any given set of* $n(n+3)/2$ *points. In particular, there is a conic passing through five points and a cubic passing through given nine points.*

We give some applications of Bezout's theorem.

THEOREM 5.10. *Suppose two cubics defined by homogeneous polynomials* F_1 *and* F_2 *have no component in common. If another cubic defined by homogeneous F passes through eight of their nine points of intersection, it passes through the ninth point also.*

PROOF. It is enough to show that $F = c_1F_1 + c_2F_2$, for two constants c_1, c_2. Suppose $F \neq c_1F_1 + c_2F_2$ for any c_1, c_2. We will show that this leads to a contradiction. Obviously, $F_1 \neq cF_2$ (otherwise, they have every component in common). Given any two points A, B we can choose (by solving two linear equations in three variables) constants c, c_1, c_2 such that the curve

$$F^* = F_{A,B} = cF - c_1F_1 - c_2F_2 = 0$$

passes through A, B and $1 \leq \deg F^* \leq 3$. If $P_1, \ldots, P_8$ are the eight points common to the three cubics, $F_{A,B} = 0$ passes through $P_1, \ldots, P_8$, as well as A and B.

Now at most three of $P_1, \ldots, P_8$ can be on a line; otherwise this line will be a common component of $F_1 = 0$ and $F_2 = 0$. Similarly, at most six of these points lie on a conic. Out of $P_1, \ldots, P_8$ two, say P_1 and P_2, always lie on a line L and five, say $P_4, \ldots, P_8$, lie on a conic C. There are three cases to be considered:

1. P_3 lies on L;
2. P_3 lies on C;
3. P_3 lies neither on L nor on C.

Case 1. Let A ($\neq P_j, j = 1, 2, 3$) be a point on L and B a point neither on L nor on C. Because L and $F^* = 0$ have four points P_1, P_2, P_3 and A is common, L is a component of $F^* = 0$. The other component of $F^* = 0$ must be C. Hence B cannot lie on $F_{A,B} = 0$ — a contradiction.

Case 2. Now take A ($\neq P_j, j = 3, \ldots, 8$) on C and B neither on L nor on C. Then $F^* = 0$ and C intersect in more than six points, so they must have a common component which has to be C. The other component of $F^* = 0$ must be L. So $F^* = 0$ does not pass through B, a contradiction.

Case 3. Taking both A, B on L, we can show that P_3 is not on $F_{A,B} = 0$ — again a contradiction. □

5.8. The Genus of a Curve

As expected, the curves get more complicated to study as their degree increases. But it is really the genus that tells us how complicated a curve is at least as far as the study of its rational solutions is concerned. Unless stated otherwise, we assume our curves to be projective and defined by irreducible polynomials with coefficients in $\mathbb{Q}$.

THEOREM 5.11*. *An irreducible curve C with singularities of order > 2 is birationally equivalent to one that has only double points as its singularities.*

As an illustration, a curve with a triple point can be transformed into one with three double points (cf. Fig. 5.7).

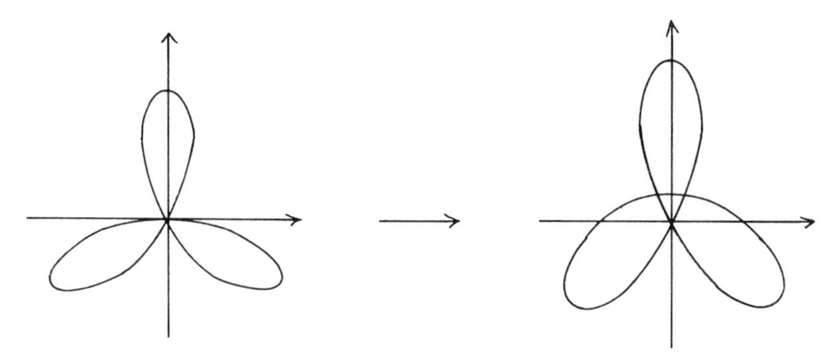

Figure 5.7. Transforming a triple point (left) into double points (right).

THEOREM 5.12. *Let C_1 be an irreducible curve of order n with m double points as its only singularities. Then*

$$g = g(C_1) = \frac{(n-1)(n-2)}{2} - m \tag{5.11}$$

is a non-negative integer.

PROOF. If $g < 0$, then the number m of double points is at least $(n-1)(n-2)/2 + 1$. Since the lines and the irreducible conics have no singularities, $n > 2$. We know that there is always a curve of degree N passing through $N(N+3)/2$ points. If $N = n - 2$, then

$$\frac{N(N+3)}{2} = \frac{(n-2)(n+1)}{2} = \frac{(n-1)(n-2)}{2} + n - 2.$$

Let C_2 be a curve of degree $n - 2$ passing through $(n-1)(n-2)/2 + 1$ double points of C_1 and further $n - 3$ points of C_1. Since each double point of a curve is counted at least twice in its intersection with another curve, C_1 and C_2 intersect in at least

$$2\left[\frac{(n-1)(n-2)}{2} + 1\right] + n - 3 = n(n-2) + 1$$

points, which is impossible by Bezout's theorem. □

THEOREM 5.13*. *Let C be an irreducible curve that is birationally equivalent to two curves C_1 and C_2 each with double points as their only singularities. Then $g(C_1) = g(C_2)$.*

The common value $g(C) = g(C_1) = g(C_2)$ is called the *genus* of the (irreducible) curve C. Note that g is a non-negative integer.

COROLLARY 5.14. *The genus of a curve is invariant under birational equivalence.*

REMARK 5.15. Our definition of the genus is geometric. There are other ways to define the genus. For the arithmetic definition involving the Riemann–Roch theorem see Ref. 2 or 6, whereas for the topological definition of the genus as the number of handles on the corresponding manifold, and the equivalence of these definitions, Ref. 4 may be consulted.

EXAMPLES 5.16.

1. Let L be a line. We have $n = 1$ and $m = 0$. Therefore,

$$g(L) = \frac{(n-1)(n-2)}{2} - m = 0.$$

2. If C is an irreducible conic, then it is of order 2. Because an irreducible conic has no singular points, $m = 0$ and hence

$$g = \frac{(n-1)(n-2)}{2} - m = 0.$$

3. Let C be the cubic $y^2 = x^3 + x^2$. Then C has only one double point as its singularity, so that the genus $g = 0$.

4. We have seen that the curve E defined by $y^2 = x^3 - x$ is nonsingular. Its genus $g = 1$.

5. The Fermat curve F: $X^n = Y^n = Z^n$ $(n \geq 3)$ has no singularity, because

$$\frac{\partial F}{\partial X} = \frac{\partial F}{\partial Y} = \frac{\partial F}{\partial Z} = 0$$

has no solution in $\mathbb{P}^2(\mathbb{C})$. Hence the genus $g = (n-1)(n-2)/2$.

THEOREM 5.17. *Let* $f(x) = x^3 + Ax + B \in \mathbb{Q}[x]$. *Then the curve*

$$E\colon y^2 = f(x) \tag{5.12}$$

has genus one $\Leftrightarrow$ *E is nonsingular* $\Leftrightarrow \Delta(f) = -4A^3 - 27B^2 \neq 0 \Leftrightarrow f(x)$ *has distinct roots.*

PROOF. It is clear from (5.11) that $g(E) = 1 \Leftrightarrow E$ is nonsingular. We have also seen (Corollary 4.10) that $\Delta(f) = 0 \Leftrightarrow f(x)$ has multiple roots. Thus it suffices to prove that E is singular $\Leftrightarrow \Delta(f) = 0$.

If $F(X, Y, Z) = 0$ is the projective model of (5.12), then $\partial F/\partial Z$ does not vanish at the point at infinity on E, i.e., at $(0, 1, 0)$. Therefore,

$$E \text{ is singular} \Leftrightarrow \frac{\partial f}{\partial x} = f'(x) \quad \text{and} \quad \frac{\partial f}{\partial y} = 2y \tag{5.13}$$

vanish simultaneously at a point on the affine part of $E \Leftrightarrow f(x)$ and $f'(x)$ have a common root $\Leftrightarrow R(f, f') = 0 \Leftrightarrow \Delta(f) = 0$. □

A straight line

$$ax + by + c = 0 \tag{5.14}$$

defined over $\mathbb{Q}$ has infinitely many rational points. (Take x to be any rational number and solve (5.14) for y.) However, a conic defined over $\mathbb{Q}$ may not have any rational point on it (e.g., $x^2 + y^2 + 1 = 0$) or it may have just one rational point, as in the case of $x^2 + y^2 = 0$. We shall exclude such conics from our discussion of the rational points on algebraic curves. Under this exclusion (which we shall assume henceforth without stating it explicitly), the straight lines and conics are rational curves and thus can be parametrized as

$$\begin{aligned} x &= \phi_1(t), \\ y &= \phi_2(t), \end{aligned} \tag{5.15}$$

by a single variable t. Since $\phi_1(t), \phi_2(t) \in \mathbb{Q}(t)$, the rational points on (5.15) are essentially given as $P(t) = (\phi_1(t), \phi_2(t))$ by varying t over the rationals. Note that the straight lines and conics are curves of genus zero. A theorem of Hilbert and Hurwitz (cf. Ref. 3) says that any curve of genus zero is birationally equivalent to either a line or a conic and thus is a rational curve. A rational curve always has infinitely many rational points, and with the exception of finitely many, these rational points are in a one-to-one correspondence with the rational numbers. Once we have put such a curve in the form (5.15), the problem is completely solved.

The curves of genus one are much harder to deal with. But thanks to Poincaré [5] such curves can be reduced to the form (5.12) with $\Delta(f) \neq 0$. It was conjectured in 1922 by Mordell and proved recently by Faltings [1] that any curve of genus larger than one has only finitely many rational points. A curve of genus one may or may not have infinitely many rational solutions and it is still an open question how to decide whether or not such a curve has only finitely many solutions. Thus in a sense the curves of genus one are the most interesting ones and the rest of the book will be devoted to their study. We state two theorems without proof.

THEOREM 5.18* (*Hilbert–Hurwitz*). *Any curve of genus zero and of degree $n \geq 3$ is birationally equivalent to a curve of degree $n - 2$.*

THEOREM 5.19* (*Poincaré*). *Any curve of genus one with a rational point is birationally equivalent to a cubic.*

Further simplification is provided by the following theorem and its corollary.

THEOREM 5.20. *Any nonsingular cubic with a rational point on it is birationally equivalent to the curve*

$$Y^2 = f(X), \tag{5.16}$$

where $\deg f(X) = 3$. *Moreover,* $\Delta(f) \neq 0$.

PROOF (*Nagell*). If P is a rational point on the cubic

$$g(x, y) = 0, \tag{5.17}$$

the tangent at P is a rational line. It meets the curve at the rational point P twice and hence its third point of intersection Q with the curve is also a rational point. Shifting the origin to Q, we may assume that

$$g(x, y) = \phi_3(x, y) + 2\phi_2(x, y) + \phi_1(x, y),$$

where ϕ_j is homogeneous of degree j. The intersection of the line $y = tx$ with the curve is given by

$$\psi_3 x^2 + 2\psi_2 x + \psi_1 = 0, \tag{5.18}$$

where $\psi_j = \phi_j(1, t)$ is a polynomial in t of degree j. Solving (5.18) we get

$$x = \frac{-\psi_2 \pm (\psi_2^2 - \psi_1\psi_3)^{1/2}}{\psi_3}. \tag{5.19}$$

Thus (5.18) is equivalent to

$$(\psi_3 x + \psi_2)^2 = \psi_2^2 - \psi_1\psi_3.$$

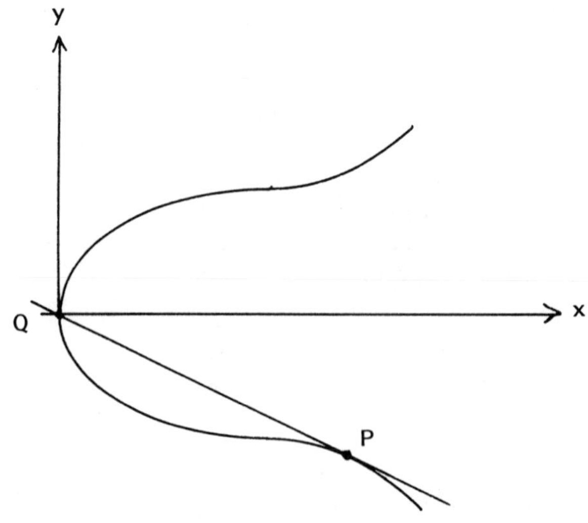

Figure 5.8. Reduction of a cubic to the standard form.

Clearly $\deg(\psi_2^2 - \psi_1\psi_3) \leq 4$. If $\deg(\psi_2^2 - \psi_1\psi_3) < 4$, we put $f = \psi_2^2 - \psi_1\psi_3$, $t = y/x = X$ and $\psi_3 x + \psi_2 = Y$. This establishes a birational correspondence between (5.16) and (5.17), because we can solve for X, Y as rational functions of x, y and vice versa. Because the genus is one, $\deg f = 3$ and $\Delta(f) \neq 0$. Now suppose that $\deg(\psi_2^2 = \psi_1\psi_3) = 4$. If $y = t_0 x$ is the tangent to the curve at P, it follows from (5.19) that t_0 is a rational root of $\psi_2^2 - \psi_1\psi_3$. We put $t = t_0 + 1/x$. Then

$$\psi_2^2 - \psi_1\psi_3 = \frac{f(x)}{x^4},$$

for a polynomial $f(x)$ with $\deg f(x) \leq 3$. In fact, $\deg(f) = 3$, because $\deg(f) < 3$ would contradict the invariance of genus. We now put $x^2(\psi_3 x + \psi_2) = Y$. As is easily seen, again we can solve for X, Y as rational functions of x, y and vice versa. □

COROLLARY 5.21. *Any curve of genus one and passing through a rational point is birationally equivalent to*

$$Y^2 = X^3 + AX + B \qquad (A, B \in \mathbb{Q})$$

with $\Delta = -4A^3 - 27B^2 \neq 0$.

PROOF. By Poincaré's theorem, any curve of genus one with a rational point is birationally equivalent to a nonsingular cubic which by Theorem 5.20 can be put in the form (5.16).

If $f(X) = aX^3 + bX^2 + cX + d$ $(a \neq 0)$, we mutiply throughout by a^2 and replace aX and aY by X and Y, respectively. Thus we may assume that $a = 1$. The substitution $X \to X - b/3$ gets rid of the square term and we get the desired form. By Theorem 5.17, the genus of this curve is one if and only if $\Delta \neq 0$. □

As a supplement we state and prove the following theorem:

THEOREM 5.22. *If a quartic curve C*

$$y^2 = ax^4 + bx^3 + cx^2 + dx + e \qquad (a \neq 0) \tag{5.20}$$

has a rational point, it is equivalent to a curve

$$Y^2 = f(X), \tag{5.21}$$

where $f(X)$ is a polynomial of degree ≤ 3. Moreover, $\deg(f(X)) = 3 \Leftrightarrow$ *the genus $g(C) = 1$.*

PROOF. (*Mordell*). (Or see the last paragraph in the proof of Theorem 5.20.) If the rational point on (5.20) is (x_0, y_0), changing x to $X - x_0$, we may assume that $(0, y_0)$ is the given rational point on (5.20), hence e is a perfect square. Therefore, by changing x to $1/X$ and y to Y/X^2, we assume that a is a perfect square. If $a = 0$, we are done. Otherwise, replacing x by

$X/\sqrt{a}$, y by $Y/\sqrt{a}$, we assume that $a = 1$. Further, the substitution $x = X - b/4$ gets rid of the term bx^3 in (5.20). Consequently, (5.20) can be written as

$$y^2 = x^4 - 6cx^2 + 4dx + e. \tag{5.22}$$

Now the substitution

$$x = \frac{1}{2}\frac{Y - d}{X - c}, \qquad y = -x^2 + (2X + c) \tag{5.23}$$

establishes a birational correspondence between the curves (5.22) and a curve of the type (5.21). To see this, first substitute in (5.22) for y from (5.23) to get

$$-2x^2(2X + c) + (2X + c)^2 = -6cx^2 + 4dx + e$$

or

$$x^2(X - c) + dx = g(X), \tag{5.24}$$

where $\deg g(X) \leq 2$. Then substitute for x from (5.23) in (5.24) to get

$$(Y - d)^2 + 2d(Y - d) = 4(X - c)g(X)$$

which gives (5.21) with $f(X) = 4(X - c)g(X) + d^2$.

5.9. Elliptic Curves

We want to study the question: *when does a (plane algebraic) curve have infinitely many rational points*? As previously mentioned, a curve of genus zero has, excluding certain trivial cases, infinitely many rational points, while according to Faltings [1] there are only finitely many rational points on a curve of genus larger than one. However, given an arbitrary curve of genus one, no one knows whether it has infinitely many rational points or not. (Both the cases do occur.) The study of elliptic curves, i.e., the curves of genus one, is one of the richest and most active fields of mathematics. An elliptic curve E may be defined to be a (projective) curve of genus one with at least one rational point. From our previous discussion, it follows that E is given by a cubic, which may be taken in the *Weierstrass form*

$$y^2 = x^3 + Ax + B, \tag{5.25}$$

where the discriminant $\Delta = \Delta(f)$ of the polynomial $f(x) = x^3 + Ax + B$ satisfies

$$\Delta = -4A^3 - 27B^2 \neq 0. \tag{5.26}$$

To be precise, we make the following definition.

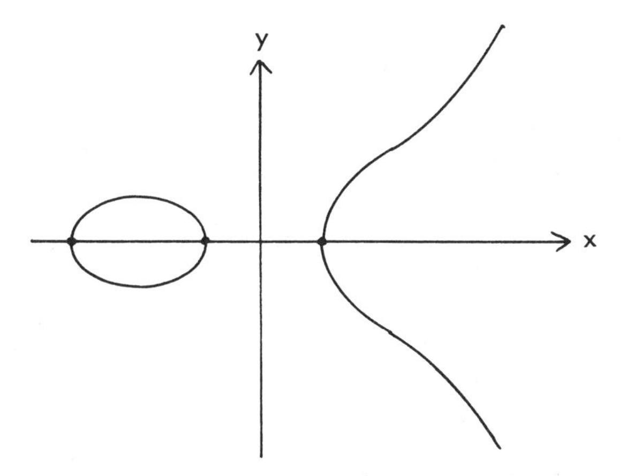

Figure 5.9. Elliptic curve in the case of three real roots.

DEFINITION 5.23. An *elliptic curve E defined over* a subfield k of $\mathbb{C}$ and written as E/k is a projective curve given by

$$Y^2Z = X^3 + AXZ^2 + BZ^3 \qquad (A, B \in k), \tag{5.27}$$

where the quantity $\Delta = -4A^3 - 27B^2$, called the *discriminant* of E, is nonzero.

There is only one point at infinity on the projective curve (5.27). It is given by $Z = 0$ and is $O = (0, 1, 0)$. It is regarded as a rational point and we may think of E as the affine curve (5.25) together with the point $O = (0, 1, 0)$ at infinity.

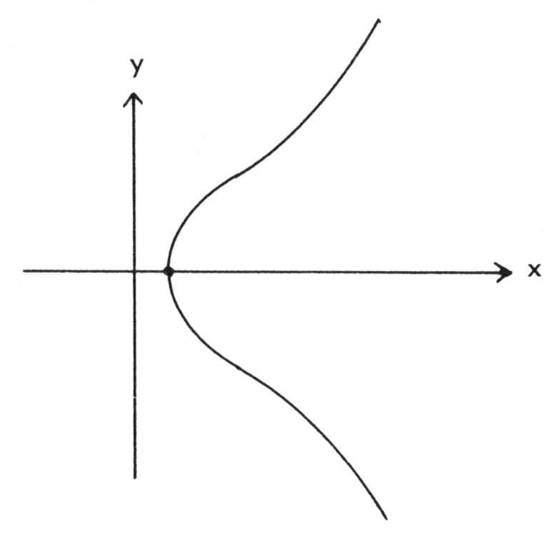

Figure 5.10. Elliptic curve in the case of one real root.

If K is any field with $k \subseteq K \subseteq \mathbb{C}$, we denote the set of all solutions of (5.27) in $\mathbb{P}^2(K)$, i.e., all solutions of (5.25) in K^2 and the point at infinity, by $E(K)$. When $K = \mathbb{C}$, $E(K)$ will be denoted by E itself. It is easily seen that $E(\mathbb{R})$ looks as in Fig. 5.9 or Fig. 5.10, according as the number of real roots of $f(x) = x^3 + Ax + B$ is three or one.

5.10. The Group Law

We define a binary operation on E as shown in Fig. 5.11. If P, Q are two points on E, the line through P, Q (the tangent at P, if $P = Q$) intersects E in a third point, which we denote by PQ. If $PQ = (x, y)$, the sum $P + Q$ is defined to be the reflection $(x, -y)$ of PQ in the x-axis. The sum $O + O$ is defined to be O. [By joining O to a point P on the affine part of E, we mean drawing a vertical line through P. The three points of intersection of a vertical line with E are (x, y), $(x, -y)$, and O. The reflection of O in the x-axis is O itself.]

This makes E into an abelian group with O as the identity. (If there are several elliptic curves under discussion, we write O_E for the identity of E.) The inverse of $P = (x, y)$ is $(x, -y)$. The only axiom to check is the associativity, i.e., for any three points P_i $(i = 1, 2, 3)$ on E,

$$(P_1 + P_2) + P_3 = P_1 + (P_2 + P_3).$$

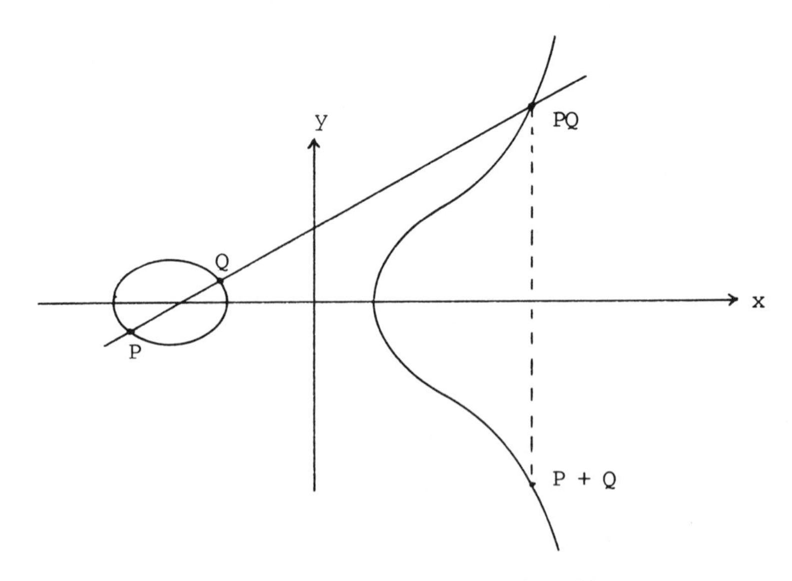

Figure 5.11. Addition on a nonsingular cubic.

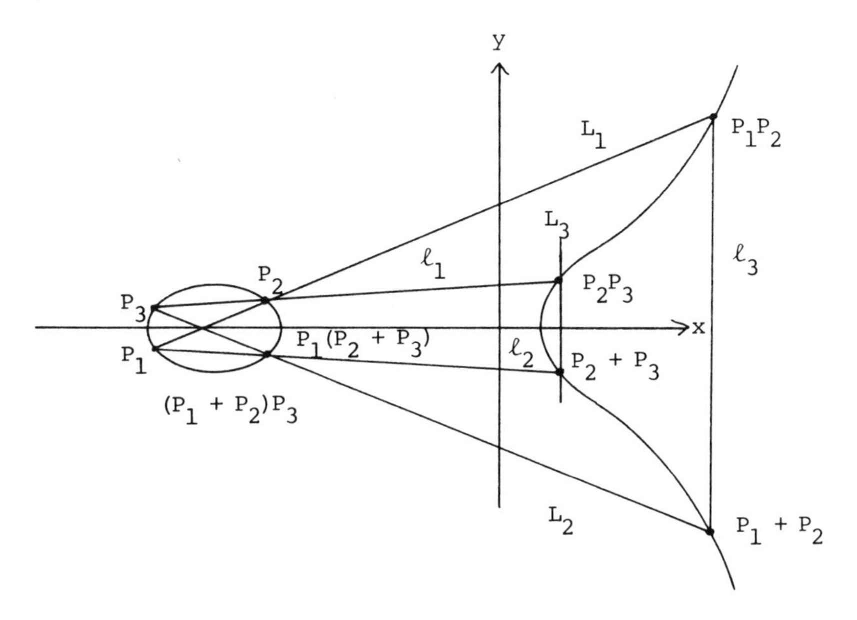

Figure 5.12. Associativity of the addition.

It is enough to prove that $P_1(P_2 + P_3) = (P_1 + P_2)P_3$. Let C_1 (or, respectively, C_2) be the cubic consisting of the three lines L_1, L_2, L_3 (respectively l_1, l_2, l_3) as shown in Fig. 5.12. Then E passes through the eight common points

$$P_1, \quad P_2, \quad P_3, \quad P_1P_2, \quad P_2P_3, \quad P_1 + P_2, \quad P_2 + P_3, \quad \text{and } O$$

of C_1 and C_2. Thus by Theorem 5.10, $P_1(P_2 + P_3)$ and $(P_1 + P_2)P_3$ must coincide with their ninth point of intersection with E.

THEOREM 5.24. *If E is defined over k and $k \subseteq K \subseteq \mathbb{C}$, the set $E(K)$ is a subgroup of E, called the group of K-rational points on E.*

PROOF. All we have to show is that if $P_1 = (x_1, y_1)$ and $P_2 = (x_2, y_2) \in E(K)$, then $P_1P_2 \in E(K)$. We may assume that these two points lie in the affine part of E and $P_1 \neq -P_2$. To determine the coordinates of P_1P_2, we consider two cases.

(1) If $P_1 \neq P_2$, the slope of the line through P_1 and P_2 is

$$m = \frac{y_1 - y_2}{x_1 - x_2}. \tag{5.28}$$

(2) For the *duplication* $2P$ of $P = (x, y)$, we have $P = P_1 = P_2$. Assume $y \neq 0$; otherwise $2P = O$ and P is a point of order 2. The slope of the tangent to (5.25) at P is now

$$m = \frac{3x^2 + A}{2y}. \tag{5.29}$$

In any case the line joining P_1 and P_2 can be written as

$$y = mx + b \qquad (m, b \in K), \tag{5.30}$$

where

$$b = y_1 - mx_1. \tag{5.31}$$

Substituting (5.30) in (5.25), we obtain

$$x^3 - m^2x^2 + (A - 2mb)x + B - b^2 = 0. \tag{5.32}$$

If $P_1P_2 = (x_3, y_3)$, then x_1, x_2, x_3 are the three roots of (5.32) and therefore

$$x_1 + x_2 + x_3 = m^2.$$

This shows that $x_3 = m^2 - (x_1 + x_2)$ is in K. That $y_3 \in K$ now follows from (5.30).

We shall denote the x (respectively y) coordinate of a point P by $x(P)$ [respectively, $y(P)$]. We have seen that

$$x(P_1 + P_2) = \left(\frac{y(P_1) - y(P_2)}{x(P_1) - x(P_2)}\right)^2 - [x(P_1) + x(P_2)], \quad \text{if } P_1 \neq \pm P_2$$

and (5.33)

$$x(2P) = \left\{\frac{3[x(P)]^2 + A}{2y(P)}\right\}^2 - 2x(P), \qquad \text{if } y(P) \neq 0.$$

REMARKS 5.25.

1. The coordinates of $P_1 \pm P_2$ are rational functions of the coordinates of P_1 and P_2.

2. Suppose $p \neq 2, 3$ is a prime and A, $B \in \mathbb{F}_p$, such that $\Delta = -4A^3 - 27B^2 \in \mathbb{F}_p^\times$. Then

$$y^2 = x^3 + Ax + B(A, B \in \mathbb{F}_p) \tag{5.34}$$

defines an *elliptic curve E over $\mathbb{F}_p$*. For any field K containing $\mathbb{F}_p$, the set of *K-rational points* on E, i.e., the set of solutions of (5.34) with x, y in K together with a *point O at infinity* is a group under the addition defined purely algebraically by the equations (5.28)-(5.33). Note that if $p = 2$, the second part of fomula (5.33) breaks down.

3. *Points of order two.* If $P = (x, y)$ on E is a point of order 2, then $P = -P = (x, -y)$, so that $y = 0$. If a_1, a_2, a_3 are the three roots of $f(x) = x^3 + Ax + B$, $P_i = (a_i, 0)$ are the three points of order 2. Together with O they form an abelian group isomorphic to $\mathbb{Z}/2\mathbb{Z} \times \mathbb{Z}/2\mathbb{Z}$. In fact, if i, j, k

are all distinct, it is clear that $P_i + P_j = P_k$. Thus for a subfield K of $\mathbb{C}$, the subgroup of $E(K)$ of points of order 2 together with O is isomorphic to

$$\begin{cases} \mathbb{Z}/2\mathbb{Z} \times \mathbb{Z}/2\mathbb{Z} & \text{if all } a_i \in K, \\ \mathbb{Z}/2\mathbb{Z} & \text{if only one } a_i \in K, \\ \{0\} & \text{otherwise.} \end{cases}$$

4. *Points of order N.* Let $E = E(\mathbb{C})$. Then the group $E[N]$ of points whose order divides $N \geq 1$ is isomorphic to $\mathbb{Z}/N\mathbb{Z} \times \mathbb{Z}/N\mathbb{Z}$ (see Section A.7).

EXAMPLES 5.26.

1. Let E be defined by

$$y^2 = x^3 - x + 1.$$

Since $\Delta = -4A^3 - 27B^2 = -23 \neq 0$, E is an elliptic curve. The point $P = (1, 1)$ is a rational point on E. Using formulas (5.28)-(5.33), we can see that $2P = (-1, 1)$, $3P = (0, -1)$, $4P = (3, -5)$, $5P = (5, 11)$, and $6P = (\frac{1}{4}, \frac{7}{8})$, etc. We shall see later (Section 7.7) that the points mP, $m = 1, 2, 3, \ldots$, are all distinct, therefore E has infinitely many rational points.

2. Let E be defined by

$$Y^2 = X^3 - 432.$$

We have seen that E is birationally equivalent to

$$E_1\colon x^3 + y^3 = 1.$$

A birational correspondence is given by

$$\begin{aligned} x &= \frac{6}{X} + \frac{Y}{6X}, & y &= \frac{6}{X} - \frac{Y}{6X}, \\ X &= \frac{12}{x + y}, & Y &= 36\,\frac{x - y}{x + y}. \end{aligned} \tag{5.35}$$

If (x, y) is a rational point on E_1, $x \neq -y$. Also if (X, Y) is a rational point on E, $X \neq 0$. Thus (5.35) gives a one-to-one and onto map from $E(\mathbb{Q})$ to $E_1(\mathbb{Q})$. By Fermat's last theorem, the only rational points on the affine part of E_1 are $(1, 0)$ and $(0, 1)$. Thus the only rational points of E are O, $(12, 36)$, $(12, -36)$, and therefore

$$E(\mathbb{Q}) \cong \mathbb{Z}/3\mathbb{Z}.$$

EXERCISES 5.27.

1. Show that $(0, 1)$ is a point of order 4 on $y^2 = x^3 - 2x + 1$. [*Hint*: $y(2P) = 0$.]

2. Show that $P = (3, 8)$ is a point of order 7 on $y^2 = x^3 - 43x + 166$. [*Hint*: Show that $8P = P$.]

References

1. G. Faltings, Endlichkeitssätze abelsche varietäten über Zahlkörpern, *Invent. Math.* **73**, 349-366 (1983).
2. W. Fulton, *Algebraic Curves*, W. A. Benjamin, New York, (1969).
3. D. Hilbert and A. Hurwitz, Über die diophantischen Gleichungen von Geschlecht Null, *Acta Math.* **14**, 217-224 (1890).
4. D. Mumford, *Algebraic Geometry I: Complex Projective Varieties*, Springer Verlag, Berlin (1976).
5. H. Poincaré, Sur les propriétés arithmétiques des courbes algébriques, *J. Math. Pures Appl.* **71**, 161-233 (1901).
6. W. M. Schmidt, *Lectures on Equations over Finite Fields: An Elementary Approach*, Part II at the University of Colorado, Boulder, 1974-1975 (unpublished).
7. J. G. Semple and L. Roth, *Algebraic geometry*, Oxford Univ. Press (1949).
8. J. Tate, Arithmetic of elliptic curves, Phillips Lectures at Haverford College 1961 (unpublished).
9. R. J. Walker, *Algebraic Curves*, Princeton Univ. Press, Princeton, New Jersey (1950).

6

The Mordell–Weil Theorem

6.1. Introduction

In 1901 Poincaré conjectured (or, rather, had tacitly assumed) [6] that all the rational points on any elliptic curve defined over $\mathbb{Q}$ are obtained from only finitely many by adding them in all possible ways. This was proved by Mordell in 1922 [3]. Soon afterwards in 1928 Weil, in his thesis [7], extended this result to the case of an "abelian variety" defined over a number field K. An *abelian variety X defined over K* is, roughly speaking, the set of common zeros in projective space of a finite number of homogeneous polynomial equations in several variables with coefficients in K, together with an abelian group law giving the coordinates of $P_1 \pm P_2$ as rational functions of the coordinates of P_1 and P_2 [5]. Weil proved that the group $X(K)$ of the K-rational points of an abelian variety X, i.e., the solutions of these polynomial equations with coordinates in K, is finitely generated. For details see Ref. 2 or 7. Later on he also gave a simpler proof, using the concepts he had introduced in his thesis, for the special case of elliptic curves. It is this proof that we shall be following (cf. Ref. 4 or 8). A very interesting account is in Cassels [1].

Let E be the elliptic curve defined by

$$y^2 = x^3 + Ax + B \qquad (A, B \in \mathbb{Q}). \tag{6.1}$$

The substitution

$$x = \frac{X}{c^2}, \qquad y = \frac{Y}{c^3} \qquad (c \in \mathbb{Q}^\times)$$

is not only a birational correspondence between (6.1) and

$$E'\colon\ Y^2 = X^3 + c^4AX + c^6B, \tag{6.2}$$

but is a group isomorphism between these two elliptic curves. In particular, $E(\mathbb{Q}) \cong E'(\mathbb{Q})$. For this substitution takes straight lines into straight lines and thus any three points are collinear on (6.1) if and only if the corresponding three points are collinear on (6.2). Actually, the converse is also true, though we will not need it here; that is, for any "$\mathbb{Q}$-isomorphism" $\phi: E_1 \to E_2$ of two elliptic curves E_i, defined by

$$y^2 = x^3 + A_i x + B_i \qquad (A_i, B_i \in \mathbb{Q}, i = 1, 2)$$

there is a constant $c \in \mathbb{Q}^\times$, such that $A_2 = c^4 A_1$, $B_2 = c^6 B_1$ and

$$\phi(x, y) = (c^2 x, c^3 y).$$

In particular, there are infinitely many nonisomorphic elliptic curves. So without loss of generality we will assume from now on that $A, B \in \mathbb{Z}$.

Because the curve (6.1) is of genus one, all the roots say a_1, a_2, a_3 of $f(x) = x^3 + Ax + B$ are distinct. We shall assume that all a_i's are in $\mathbb{Q}$. Otherwise, we can work in a number field containing these roots of $f(x)$ and the proof is essentially the same. Note that these roots of $f(x)$ are, in fact, in $\mathbb{Z}$. For, if $a = s/t$ with $(s, t) = 1$ and $f(a) = 0$, then

$$\frac{s}{t}\left(\frac{s^2 + At^2}{t^2}\right) = -B \in \mathbb{Z}$$

and this can happen with $(s, t) = 1$ only if $t = 1$. (See also Exercise 4.24.)

6.2. Heights of Rational Points

For $x = s/t \in \mathbb{Q}^\times$ with $(s, t) = 1$, we define the *height* $H(x)$ of x by $H(x) = \max(|s|, |t|)$. We put $H(0) = 1$. If $P = (x, y)$ is a point with rational coordinates, the *height* $H(P)$ of P is defined to be the positive integer $H(x)$. Clearly, for any real number $c > 0$, there are only finitely many rational points on elliptic curve E with $H(P) \leq c$. The group $E(\mathbb{Q})$ will be finitely generated if any point P of $E(\mathbb{Q})$ is a sum of points of $E(\mathbb{Q})$ of bounded heights, this bound $c = c(E)$ depending only on E. This is the main idea of the proof. We make few observations.

THEOREM 6.1. *If $P = (x, y) \in E(\mathbb{Q})$, the group of $\mathbb{Q}$-rational points of an elliptic curve E defined by*

$$y^2 = x^3 + Ax + B \qquad (A, B \in \mathbb{Z}), \tag{6.3}$$

then $x = s/t^2$, $y = u/t^3$ with $(s, t) = (u, t) = 1$ and $t \geq 1$.

PROOF. Let $x = s/S$, $y = u/U$ with $(s, S) = (u, U) = 1$ and $S, U \geq 1$. Then

$$u^2 S^3 = U^2 s^3 + AsU^2 S^2 + BU^2 S^3. \tag{6.4}$$

We shall show that $S^3 = U^2$, because then $S = t^2$ and $U = t^3$.

From (6.4) we see that $U^2|u^2S^3$, but $(u, U) = 1$, so $U^2|S^3$. From (6.4) again, $S^2|s^3U^2$. But $(s, S) = 1$, so $S^2|U^2$. This together with (6.4) shows that $S^3|s^3U^2$ which shows that $S^3|U^2$. Since $S, U \geq 1$, $S^3 = U^2$. □

THEOREM 6.2. *If E is defined by (6.3), put $c = c(E) = (1 + |A| + |B|)^{1/2}$. Then for any $P = (x, y)$ in $E(\mathbb{Q})$,*

$$H(y) \leq c(H(P))^{3/2}.$$

PROOF. We know that $P = (s/t^2, u/t^3)$ with $(s, t) = (u, t) = 1$ and $t \geq 1$. Putting in (6.3),

$$u^2 = s^3 + Ast^4 + Bt^6.$$

Since $|s| \leq H(x)$ and $t^2 \leq H(x)$, we have

$$u^2 \leq (1 + |A| + |B|)H(x)^3.$$

Therefore

$$H(y) \leq \max(|u|, t^3) \leq cH(P)^{3/2}.$$ □

NOTATION 6.3. Let $f, g: S \to \mathbb{R}^+$ be two functions from a set S into the set of positive real numbers $\mathbb{R}^+$. We shall use the symbol $f(x) = O(g(x))$ on S to mean that there is a constant $c > 0$, depending on f and g only, such that $f(x) \leq cg(x)$ for all x in S. In this notation, the above theorem can be stated as $H(y(P)) = O(H(P)^{3/2})$ on $E(\mathbb{Q})$. Clearly, if $f = O(g)$ and $g = O(h)$, then $f = O(h)$. Also note that a finite subset S_1 of S can always be ignored, i.e., $f = O(g)$ on $S \Leftrightarrow f = O(g)$ on $S - S_1 = \{x \in S | x \text{ is not in } S_1\}$.

6.3. Abscissas of Collinear Points

There are special expressions for $x(P_1 + P_2)$ as rational functions of $x(P_1)$ and $x(P_2)$ involving the three roots a_1, a_2, a_3 of $f(x)$ which play an important role in Weil's proof of Mordell's theorem (cf. Ref. 8). Let $P_i = (x_i, y_i) \in E(\mathbb{Q})$, $i = 1, 2$.

(1) First let $P_1 \neq \pm P_2$. To calculate $x_3 = x(P_1 + P_2)$ we intersect the line

$$y = y_1 + \frac{y_2 - y_1}{x_2 - x_1}(x - x_1)$$

through P_1 and P_2 with (6.3) and solve the resulting equation

$$\left[y_1 + \frac{y_2 - y_1}{x_2 - x_1}(x - x_1)\right]^2 = x^3 + Ax + B \tag{6.5}$$

for the three values x_1, x_2, and x_3 of x. If $a = a_1, a_2$, or a_3, we put $x = X + a$, $x_1 = X_1 + a$ and $x_2 = X_2 + a$ in (6.5). Because $f(a) = 0$, $X = 0$ is a root of $f(X + a)$ and so (6.5) becomes

$$\left[y_1 + \frac{y_2 - y_1}{X_2 - X_1}(X - X_1)\right]^2 = X^3 + C_2X^2 + C_1X. \tag{6.6}$$

The constant term in (6.6) is the product of its three solutions $X = x_1 - a$, $x_2 - a$, $x_3 - a$ and we have

$$(x_1 - a)(x_2 - a)(x_3 - a) = \left(y_1 - \frac{y_2 - y_1}{x_2 - x_1}X_1\right)^2. \tag{6.7}$$

Since $x(P_1 + P_2) = x_3$, replacing X_1 with $x_1 - a$, this takes the form

$$x(P_1 + P_2) - a = \frac{1}{(x_1 - a)(x_2 - a)}\left[\frac{y_1(x_2 - a) - y_2(x_1 - a)}{x_2 - x_1}\right]^2. \tag{6.8}$$

(2) The x-coordinates of the third point of intersection of the line through P_1 and P_2 with (6.3) is also given by

$$x_3 = \left(\frac{y_2 - y_1}{x_2 - x_1}\right)^2 - (x_2 + x_1).$$

Using the fact that P_1 and P_2 are on (6.3), this can be written as

$$x(P_1 + P_2) = \frac{(x_1 + x_2)(x_1x_2 + A) + 2B - 2y_1y_2}{(x_2 - x_1)^2}. \tag{6.9}$$

(3) For the duplication $2P$ of a point $P = (x, y)$ with $y \neq 0$, the equation (6.7) is still valid except for the slope $(y_2 - y_1)/(x_2 - x_1)$ of the line joining P_1 and P_2, we now use the slope of the tangent to (6.3) at P to get

$$\begin{aligned} x(2P) - a &= \frac{1}{(x - a)^2}\left[y - \frac{3x^2 + A}{2y}(x - a)\right]^2 \\ &= \frac{1}{(x - a)^2}\left[\frac{2y^2 - (3x^2 + A)(x - a)}{2y}\right]^2. \end{aligned}$$

Because $x = a$ is a root of $f(x) = x^3 + Ax + B$, one can see, using the division algorithm, that

$$y^2 = (x - a)(x^2 + ax + A + a^2),$$

so

$$2y^2 - (3x^2 + A)(x - a) = (x - a)(-x^2 + A + 2ax + 2a^2).$$

Therefore, for each $j = 1, 2, 3$, we have

$$x(2P) - a_j = \left(\frac{-x^2 + A + 2a_jx + 2a_j^2}{2y}\right)^2. \tag{6.10}$$

6.4. Review of Linear Algebra

The determinant D of the *van der Monde matrix*

$$M = \begin{pmatrix} 1 & a_1 & a_1^2 \\ 1 & a_2 & a_2^2 \\ 1 & a_3 & a_3^2 \end{pmatrix}$$

is given by

$$D = \prod_{i>j} (a_i - a_j).$$

Suppose $D \neq 0$, i.e., a_1, a_2, a_3 are all distinct. Further assume that a_1, a_2, a_3 are all in $\mathbb{Z}$. Then D is a nonzero integer and for α_1, α_2, α_3 in $\mathbb{Q}$, the system of linear equations $M\mathbf{u} = \boldsymbol{\alpha}$, where

$$\mathbf{u} = \begin{pmatrix} u_1 \\ u_2 \\ u_3 \end{pmatrix} \quad \text{and} \quad \boldsymbol{\alpha} = \begin{pmatrix} \alpha_1 \\ \alpha_2 \\ \alpha_3 \end{pmatrix}$$

has a unique solution $\mathbf{u} = M^{-1}\boldsymbol{\alpha}$. In fact,

$$u_i = \frac{1}{D}(c_{i1}\alpha_1 + c_{i2}\alpha_2 + c_{i3}\alpha_3), \tag{6.11}$$

where the cofactors c_{ij} of M are in $\mathbb{Z}[a_1, a_2, a_3]$. Moreover, if α_j are all integers, then so is $Du_i (i = 1, 2, 3)$.

6.5. Descent

The proof of finite generation of $E(\mathbb{Q})$ involves a descent. By subtracting a suitable point of a finite set $\{Q_1, \ldots, Q_n\}$ of points of $E(\mathbb{Q})$ from a given point P we get a point that is a multiple of a point of smaller height. Since the height of a point is a positive integer, the process terminates.

Proposition 6.4. *Given a point Q of $E(\mathbb{Q})$, there is a constant $c_1 > 0$, depending on E and Q only, such that for all P in $E(\mathbb{Q})$,*

$$H(P + Q) \leq c_1(H(P))^2.$$

Proof. Write $P = (x/t^2, y/t^3)$, $Q = (m/l^2, n/l^3)$ and $P + Q = (X/Z^2, Y/Z^3)$. Assume that $P \neq O$ or $\pm Q$. Then by (6.9),

$$\frac{X}{Z^2} = \frac{(xl^2 + mt^2)(mx + Al^2t^2) + 2Bl^4t^4 - 2nylt}{(mt^2 - nl^2)^2}.$$

Because $(X, Z) = 1$, Z^2 (or, respectively, $|X|$) is less than the absolute value of the denominator (respectively, numerator) of the right-hand side of this equation. Since $H(y(P)) = O(H(P))^{3/2}$, it follows that

$$H(P+Q) \le \max(|X|, Z^2) \le c_1 H(P)^2,$$

c_1 depending only on m, n, l, A, and B. □

PROPOSITION 6.5. *There is a constant $c_2 > 0$, such that for all P in $E(\mathbb{Q})$, $H(P) \le c_2 H(2P)^{1/4}$.*

PROOF. We assume that $P \ne O$ or $(a_j, 0)$, $j = 1, 2, 3$. Put $P = (x/t^2, y/t^3)$ and $2P = (X/Z^2, Y/Z^3)$. From (6.10), we see that for $j = 1, 2, 3$,

$$(X - a_j Z^2)^{1/2} = \frac{Z}{2yt}(-x^2 + At^4 + 2a_j xt^2 + 2a_j^2 t^4),$$

i.e.,

$$\alpha_j = u_1 + a_j u_2 + a_j^2 u_3,$$

where

$$\alpha_j = (X - a_j Z^2)^{1/2}$$

and

$$u_1 = \frac{-x^2 + At^4}{2yt} Z, \qquad u_2 = \frac{Zxt}{y}, \qquad u_3 = \frac{Zt^4}{yt}.$$

Since for each j, $\alpha_j \in \mathbb{Z}$, it follows from (6.11) that for $i = 1, 2, 3$, Du_i is an integer and therefore so is

$$D(Au_3 - 2u_1) = \frac{Dx^2 Z}{yt}. \tag{6.12}$$

It is clear from

$$y^2 = x^3 + Axt^4 + Bt^6, \qquad (x, t) = 1$$

that any common divisor d of x and y is a factor of B. If we write (6.12) as

$$D(Au_3 - 2u_1) = \left(\frac{x}{d}\right)^2 \cdot \frac{DZd}{(y/d)t},$$

it follows that DZd^2/yt is an integer and $(x/d)^2$ is a divisor of $D(Au_3 - 2u_1)$, so x^2 is a factor of $DB^2(Au_3 - 2u_1)$.

Finally,

$$\frac{B^2 DZ}{yt} t^4 = B^2 Du_3$$

shows that t^4 is a divisor of $B^2 Du_3$.

We have shown that x^2 (or, respectively, t^4) is a divisor of $DB^2(Au_3 - 2u_1)$ (respectively, B^2Du_3). But u_j's are fixed linear combinations of $\alpha_j = (X - a_jZ^2)^{1/2}$. Therefore, $H(P)^2 \le \max(x^2, t^4) = O(X - a_jZ^2)^{1/2}$. Because

$$(X - a_jZ^2)^{1/2} = O(H(2P))^{1/2},$$

this completes the proof. □

THEOREM 6.6. *Suppose Q is a fixed point of $E(\mathbb{Q})$. Then there is a constant $c = c(Q) > 0$, depending on Q and E only, such that*

$$H(R) \le c(H(P))^{1/2}$$

for any two points P and R of $E(\mathbb{Q})$ satisfying

$$P + Q = 2R.$$

PROOF. Apply Propositions 6.4 and 6.5:

$$\begin{aligned} H(R) &\le c_2H(2R)^{1/4} = c_2H(P+Q)^{1/4} \\ &\le c_2(c_1H(P)^2)^{1/4} = c(H(P))^{1/2}, \end{aligned}$$

with $c = c_2c_1^{1/4}$. □

6.6. The Mordell–Weil Theorem

If G is an abelian group written multiplicatively, then $G^{(n)} = \{g^n \mid g \in G\}$ is a subgroup of G for each integer $n \ge 1$. In additive notation, we write $nG = \{ng \mid g \in G\}$.

First we prove the weak Mordell–Weil theorem, which states that $E(\mathbb{Q})/2E(\mathbb{Q})$ is finite. The (strong) Mordell–Weil theorem then follows immediately.

Let $\beta: \mathbb{Q}^\times \to G = \mathbb{Q}^\times/\mathbb{Q}^{\times 2}$ be the canonical homomorphism sending each x in $\mathbb{Q}^\times$ to its coset in G. Note that G is an infinite abelian group in which each element is of order 2 and each finite subgroup of G is of order 2^s, for an integer $s \ge 0$. For $j = 1, 2, 3$ let $\phi_j: E(\mathbb{Q}) \to G$ be the function given by

$$\phi_j(P) = \begin{cases} 1, & \text{if } P = O, \\ \beta([x(P) - a_i][x(P) - a_k]) & \text{if } P = (a_j, 0), \\ \beta(x(P) - a_j) & \text{otherwise.} \end{cases}$$

[Note that if $x \ne a_1, a_2, a_3$, then $\beta((x - a_1)(x - a_2)) = \beta(x - a_3)$ because $y^2 = \prod (x - a_i)$.] We now define $\phi: E(\mathbb{Q}) \to G^3 = G \times G \times G$ by $\phi(P) = (\phi_1(P), \phi_2(P), \phi_3(P))$. In view of the formulas (6.8) and (6.10), ϕ is a group homomorphism. Moreover, ϕ has the following properties:

THEOREM 6.7. *The Image $\phi(E(\mathbb{Q}))$ of ϕ in G^3 is finite.*

PROOF. Let $P = (x/t^2, y/t^3) \in E(\mathbb{Q})$. Then

$$y^2 = (x - a_1 t^2)(x - a_2 t^2)(x - a_3 t^2).$$

If d is a common divisor of $x - a_i t^2$ and $x - a_j t^2$ $(i \neq j)$, then $d \mid (a_i - a_j)t^2$. But $(d, t) = 1$, so $d \mid a_i - a_j$. Therefore, we can write the above equation as

$$y^2 = d \frac{x - a_1 t^2}{d_1} \cdot \frac{x - a_2 t^2}{d_2} \cdot \frac{x - a_3 t^2}{d_3},$$

where $d = d_1 d_2 d_3 | D = \prod_{i>j} (a_i - a_j)$ and all the factors on the right-hand side are positive, mutually coprime, and thus are perfect squares. Consequently

$$\beta\left(\frac{x}{t^2} - a_j\right) = \beta\left(\frac{x - a_j t^2}{d_j} \cdot \frac{1}{t^2} \cdot d_j\right) = \beta(d_j).$$

So for any P in $E(\mathbb{Q})$, $\phi(P) = (\beta(d_1), \beta(d_2), \beta(d_3))$, where $d_j \mid D$. Because there are only finitely many divisors of D, we are done. □

THEOREM 6.8. $\mathrm{Ker}(\phi) = 2E(\mathbb{Q})$.

PROOF. If $P \in 2E(\mathbb{Q})$, it is obvious from formula (6.10) for the duplication of a point that $P \in \mathrm{Ker}(\phi)$. Conversely, we must show that if P is in $\mathrm{Ker}(\phi)$, then $P = 2Q$ for some Q in $E(\mathbb{Q})$. Let

$$x(P) - a_j = \alpha_j^2, \qquad j = 1, 2, 3. \tag{6.13}$$

There is a unique solution [given by (6.11)] of the linear equations

$$u_1 + a_j u_2 + a_j^2 u_3 = \alpha_j \qquad (j = 1, 2, 3) \tag{6.14}$$

in the variables u_1, u_2, u_3. Substituting (6.14) in (6.13) and using the fact that $a_j^3 + Aa_j + B = 0$, the resulting equations can be written simultaneously as

$$[u_1^2 - 2u_2u_3B - x(P)]\mathbf{v}_0 + (2u_1u_2 - 2u_2u_3A - Bu_3^2 + 1)\mathbf{v}_1 + (u_2^2 + 2u_1u_3 - Au_3^2)\mathbf{v}_2 = \mathbf{0},$$

where

$$\mathbf{v}_0 = \begin{pmatrix} 1 \\ 1 \\ 1 \end{pmatrix}, \qquad \mathbf{v}_1 = \begin{pmatrix} a_1 \\ a_2 \\ a_3 \end{pmatrix}, \quad \text{and} \quad \mathbf{v}_2 = \begin{pmatrix} a_1^2 \\ a_2^2 \\ a_3^2 \end{pmatrix}.$$

It follows from the linear independence of $\mathbf{v}_0, \mathbf{v}_1$, and $\mathbf{v}_2$ that

$$u_1^2 - 2u_2u_3B = x(P), \tag{6.15}$$

$$2u_1u_2 - 2u_2u_3A - Bu_3^2 = -1 \tag{6.16}$$

and

$$u_2^2 + 2u_1u_3 - Au_3^2 = 0. \tag{6.17}$$

From (6.17) and (6.16) we see that $u_3 \neq 0$.

Now eliminate u_1 from (6.16) and (6.17) to obtain

$$u_2^3 + Au_2u_3^2 + Bu_3^3 = u_3.$$

Since $u_3 \neq 0$, this gives

$$\left(\frac{1}{u_3}\right)^2 = \left(\frac{u_2}{u_3}\right)^3 + A\left(\frac{u_2}{u_3}\right) + B.$$

So the point $Q = (x, y)$ with

$$x = \frac{u_2}{u_3}, \qquad y = \frac{1}{u_3} \tag{6.18}$$

is in $E(\mathbb{Q})$. Dividing (6.17) throughout by u_3^2, we obtain

$$u_1 = \frac{-x^2 + A}{2y}. \tag{6.19}$$

Substituting for u_j's in terms of x, y from (6.18) and (6.19) in (6.14) we obtain

$$\alpha_j = \frac{-x^2 + A}{2y} + \frac{xa_j}{y} + \frac{a_j^2}{y}.$$

This substituted in (6.13) is the duplication formula (6.10) for $P = 2Q$. □

THEOREM 6.9 (*The weak Mordell–Weil Theorem*). *The quotient group $E(\mathbb{Q})/2E(\mathbb{Q})$ is finite.*

PROOF. By Theorem 2.34, $E(\mathbb{Q})/\mathrm{Ker}(\phi)$ is isomorphic to $\phi(E(\mathbb{Q}))$. Theorem 6.9 now follows from Theorems 6.7 and 6.8. □

COROLLARY 6.10. *The order of the quotient group $E(\mathbb{Q})/2E(\mathbb{Q})$ is 2^s, for an integer $s \geq 0$.*

Now we are in a position to prove the Mordell–Weil theorem. Note its similarity to Dirichlet's theorem (Theorem 4.32).

THEOREM 6.11 (*The Mordell–Weil Theorem*). *The group $E(\mathbb{Q})$ is finitely generated.*

PROOF. Choose a set of coset representatives $Q_1, \ldots, Q_n$ of $E(\mathbb{Q})/2E(\mathbb{Q})$ in $E(\mathbb{Q})$. Then any P in $E(\mathbb{Q})$ can be written as

$$P = Q_{i(1)} + 2P_1$$

for some P_1 in $E(\mathbb{Q})$ and $1 \le i(1) \le n$. Similarly,

$$P_1 = Q_{i(2)} + 2P_2$$
$$P_2 = Q_{i(3)} + 2P_3$$
$$\vdots$$
$$P_{m-1} = Q_{i(m)} + 2P_m.$$

Thus $P = Q_{i(1)} + 2Q_{i(2)} + 2^2 Q_{i(3)} + \cdots + 2^{m-1} Q_{i(m)} + 2^m P_m$. Let $c_i = c(-Q_i)$ be the constant appearing in Theorem 6.6 for Q replaced by $-Q_i$. Put $c = \max(c_1, \ldots, c_n)$. Then by Theorem 6.6, for $1 < j \le m$ we have $H(P_j) \le cH(P_{j-1})^{1/2}$. Therefore, if $H(P_{j-1}) > c^2$, we have

$$H(P_j) \le cH(P_{j-1})^{1/2} < H(P_{j-1}).$$

Thus $E(\mathbb{Q})$ is generated by the finite set

$$\{Q_1, \ldots, Q_n\} \cup \{P \in E(\mathbb{Q}) \mid H(P) \le c^2\}. \qquad \square$$

References

1. J. W. S. Cassels, Mordell finite basis theorem revisited, Math. Proc. Cambridge Phil. Soc. **100** 31–41 (1986).
2. Yu. I. Manin, Mordell-Weil theorem, Appendix II to Ref. 5 below.
3. L. J. Mordell, On the rational solutions of the indeterminate equations of the 3rd and the 4th degree, *Proc. Cambridge Phil. Soc.* **21** 179–192 (1922).
4. L. J. Mordell, *Diophantine Equations*, Academic, London (1969).
5. D. Mumford, *Abelian Varieties*, Oxford Univ. Press, London (1974).
6. H. Poincaré, Sur les pròpriétés arithmétiques des courbes algébriques, *J. Math. Pures Appl.* **71** 161–223, (1901).
7. A. Weil, L'arithmétique sur les courbes algébriques, *Acta Math.* **52** 281–315, (1928).
8. A. Weil, Sur un théorème de Mordell, *Bull. Sci. Math.* **54** 182–191, (1930).

7

Computation of the Mordell–Weil Group

7.1. Introduction

If G is an abelian group (written additively), the elements $g_1, \ldots, g_r$ of G are called *independent* if

$$m_1 g_1 + \cdots + m_r g_r = 0 \qquad (m_j \in \mathbb{Z})$$

is possible only with $m_1 = \cdots = m_r = 0$. Thus if one of $g_1, \ldots, g_r$ is of finite order, $g_1, \ldots, g_r$ cannot be independent. For any elliptic curve E defined over $\mathbb{Q}$ the group $E(\mathbb{Q})$ of rational points on E is finitely generated. The *(Mordell–Weil) rank* $r_{\mathbb{Q}}(E)$ of E is defined to be the maximum number of independent elements in $E(\mathbb{Q})$. In particular, $r_{\mathbb{Q}}(E) = 0$ if and only if $E(\mathbb{Q})$ is finite (consisting of points of finite order). If $r = r_{\mathbb{Q}}(E)$, then

$$E(\mathbb{Q}) \cong \underbrace{\mathbb{Z} \oplus \cdots \oplus \mathbb{Z}}_{r \text{ copies}} \times E(\mathbb{Q})_{\text{tor}}.$$

In order to know $E(\mathbb{Q})$ up to isomorphism, we need to know

1. the rank $r = r_{\mathbb{Q}}(E)$;
2. $E(\mathbb{Q})_{\text{tor}}$.

There is no general method to compute $r_{\mathbb{Q}}(E)$ for an arbitrary elliptic curve E that is known to lead to a decision in all cases. However, Fermat's method of descent often gives the answer. We illustrate this for some curves given in the form

$$E\colon y^2 = x^3 + Ax. \tag{7.1}$$

All such curves pass through the origin $\mathbf{0} = (0, 0)$. We may assume that A is an integer free of fourth powers. We follow Tate [4] (see also Ref. 1).

7.2. Factorization of the Duplication Map

We associate with E defined by (7.1) another elliptic curve

$$\bar{E}: y^2 = x^3 + \bar{A}x,$$

where $\bar{A} = -4A$. Because

$$\bar{\bar{E}}: y^2 = x^3 + 2^4 Ax,$$

we see at once that $\bar{\bar{E}} \cong E$ and this isomorphism $\psi: \bar{\bar{E}} \to E$ is given by

$$\psi(x, y) = \left(\frac{x}{2^2}, \frac{y}{2^3}\right).$$

THEOREM 7.1. *For a point $P = (x, y)$ on E, put*

$$\phi(P) = \begin{cases} O_{\bar{E}} & \text{if } P = O_E \text{ or } \mathbf{0}, \\ \left(x + \dfrac{A}{x}, \dfrac{y}{x}\left(x - \dfrac{A}{x}\right)\right) & \textit{otherwise.} \end{cases}$$

Then ϕ is a homomorphism from E into $\bar{E}$ with $\operatorname{Ker}(\phi) = \{\mathbf{0}, O\}$.

PROOF. Clearly, for $x \neq 0$,

$$\bar{x} = x(\phi(P)) = x + \frac{A}{x} = \frac{y^2}{x^2}.$$

First note that $\phi(P) = (\bar{x}, \bar{y})$ is on $\bar{E}$, for

$$\begin{aligned} \bar{y}^2 &= \frac{y^2}{x^2}\left(x - \frac{A}{x}\right)^2 \\ &= \frac{y^2}{x^2}\left[\left(x + \frac{A}{x}\right)^2 - 4A\right] \\ &= \bar{x}(\bar{x}^2 + \bar{A}). \end{aligned}$$

To show that ϕ is a homomorphism, we must show that

$$\phi(P_1 + P_2) = \phi(P_1) + \phi(P_2). \tag{7.2}$$

If one of $P_i = O$ or both $P_i = \mathbf{0}$, there is nothing to prove and we have the following two cases to consider.

Case 1. $P_1 = \mathbf{0}$, $P_2 = (x, y) \neq O$, $\mathbf{0}$. If (X, Y) is the third point of intersection with E of the line

$$Y = \frac{y}{x} X \tag{7.3}$$

joining $\mathbf{0}$ and P_2, then $\mathbf{0} + P_2 = (X, -Y)$. To prove (7.2), i.e.,

$$\phi(\mathbf{0} + P_2) = \phi(P_2),$$

first note that by (7.3),

$$x(\phi(\mathbf{0} + P_2)) = \left(\frac{Y}{X}\right)^2 = \left(\frac{y}{x}\right)^2 = x(\phi(P_2)).$$

So we must show that

$$y(\phi(\mathbf{0} + P_2)) = y(\phi(P_2)). \tag{7.4}$$

We may assume that $X \neq \pm x$, for if $X = x$, then $2P_2 = \mathbf{0}$ and $\mathbf{0} + P_2 = -P_2$. It can be checked that

$$x(2P_2) = \left(\frac{x^2 - A}{2y}\right)^2.$$

This implies that $x^2 - A = 0$ and hence

$$y(\phi(P_2)) = \frac{y}{x}\left(\frac{x^2 - A}{x}\right) = 0 = y(\phi(-P_2)) = y(\phi(\mathbf{0} + P_2)).$$

And if $X = -x$, then $y = Y = 0$ and (7.4) is obvious. Since $X \neq \pm x$, it is clear from (7.3) that (7.4) holds if and only if

$$-\left(X - \frac{A}{X}\right) = x - \frac{A}{x}$$

$$\Leftrightarrow$$

$$x + X = A\frac{x + X}{xX}$$

$$\Leftrightarrow$$

$$A = xX. \tag{7.5}$$

Because (x, y) and (X, Y) satisfy (7.1),

$$x + \frac{A}{x} = \left(\frac{y}{x}\right)^2 \quad \text{and} \quad X + \frac{A}{X} = \left(\frac{Y}{X}\right)^2.$$

These equations together with (7.3) imply that

$$x - X = A\frac{x - X}{xX}$$

which gives (7.5).

Case 2. $P_i \neq O, \mathbf{0}\,(i = 1, 2)$. By definition, $\phi(-P) = -\phi(P)$ and so we may assume that $P_1 \neq -P_2$. It is enough to show that if P_1, P_2, P_3 (no $P_i = O$

or $\mathbf{0}$) are three collinear points on E, then $\phi(P_1)$, $\phi(P_2)$, $\phi(P_3)$ are collinear points on $\bar{E}$. If P_i lie on the line

$$L: y = mx + b,$$

then $b \neq 0$; otherwise $\mathbf{0}$ will be a fourth point of intersection of E with L, a contradiction to Bezout's theorem. Also $P_1 \neq \pm P_2$ implies that L is not vertical. It can be checked that $\phi(P_i)$ all lie on the line

$$\bar{L}: y = \bar{m}x + \bar{b},$$

with

$$\bar{m} = \frac{mb - A}{b} \quad \text{and} \quad \bar{b} = \frac{b^2 + Am^2}{b}.$$

That $\operatorname{Ker}(\phi) = \{O, \mathbf{0}\}$ is obvious. □

THEOREM 7.2.

1. *For a point (X, Y) of $\bar{E}(\mathbb{Q})$ with $X \neq 0$ to be in $\phi(E(\mathbb{Q}))$ it is necessary and sufficient that $X \in \mathbb{Q}^{\times 2}$.*
2. The point $\mathbf{0} = (0, 0) \in \phi(E(\mathbb{Q}))$ *if and only if* $-A \in \mathbb{Q}^{\times 2}$.

PROOF.

1. In view of the definition of ϕ, we have only to show the sufficiency. Let $X = \omega^2$ with $\omega \in \mathbb{Q}^\times$. Put $P = (x, y)$ with

$$x = \tfrac{1}{2}(X + Y/\omega), \qquad y = \omega x.$$

Using the fact that $Y^2 + 4AX = X^3$, we get

$$\begin{aligned} x^3 + Ax &= x(x^2 + A) \\ &= x[\tfrac{1}{4}(X + Y/\omega)^2 + A] \\ &= x\frac{X^3 + 2XY\omega + Y^2 + 4AX}{4X} \\ &= xX\tfrac{1}{2}(X + Y/\omega) = Xx^2 = (\omega x)^2 = y^2. \end{aligned}$$

Since $x(\phi(P)) = X$, by changing y to $-y$, if necessary, it is obvious that $\phi(P) = (X, Y)$.

2. $\mathbf{0} \in \phi(E(\mathbb{Q}))$ if and only if for some $x \in \mathbb{Q}^\times$, $(x, 0) \in E(\mathbb{Q})$, i.e., $x^2 + A = 0$ or $-A \in \mathbb{Q}^{\times 2}$. □

Let $\psi: \bar{\bar{E}} \to E$ be the isomorphism defined at the beginning of this section and let $\bar{\phi}: \bar{E} \to \bar{\bar{E}}$ be defined in a similar way as the homomorphism $\phi: E \to \bar{E}$.

THEOREM 7.3. *The composition* $\Phi = \psi\bar{\phi}\phi \colon E \to E$ *is the duplication map, i.e., for all P in E,* $\Phi(P) = \pm 2P$.

PROOF. It is enough to show that for all $P = (x, y)$ in E, $x(\Phi(P)) = x(2P)$. By the definitions of the various functions composing Φ,

$$x(\Phi(P)) = x(\psi(\bar{\phi}(\phi(P))))$$
$$= \frac{1}{2^2}\frac{\left[\frac{y}{x}\left(x - \frac{A}{x}\right)\right]^2}{\left(\frac{y^2}{x^2}\right)^2}$$
$$= \left(\frac{x^2 - A}{2y}\right)^2.$$

On the other hand, using the fact that $y^2 = x^3 + Ax$,

$$x(2P) = \left(\frac{3x^2 + A}{2y}\right)^2 - 2x$$
$$= \frac{9x^4 + 6Ax^2 + A^2 - 8xy^2}{4y^2}$$
$$= \left(\frac{x^2 - A}{2y}\right)^2. \qquad \square$$

7.3. A Formula for the Rank

Let E be defined by (7.1) and $r = r_{\mathbb{Q}}(E)$. Since the group $E(\mathbb{Q})$ is finitely generated, by Theorem 2.32,

$$\Gamma \overset{\text{def}}{=} E(\mathbb{Q}) \cong \underbrace{\mathbb{Z} \times \cdots \times \mathbb{Z}}_{r \text{ copies}} \times \mathbb{Z}/p_1^{n_1}\mathbb{Z} \times \cdots \times \mathbb{Z}/p_k^{n_k}\mathbb{Z}.$$

If we put $G = \mathbb{Z}/p^n\mathbb{Z}$, then it is clear [Exercise 2.24(2)] that the index

$$[G \colon 2G] = \begin{cases} 2 & \text{if } p = 2, \\ 1 & \text{if } p > 2. \end{cases}$$

Therefore, if q is the number of j, such that $p_j = 2$, then

$$[\Gamma \colon 2\Gamma] = 2^r \cdot 2^q.$$

For P in $E(\mathbb{Q})_{\text{tor}}$, we write

$$P = \sum_{j=1}^{k} m_j Q_j,$$

where Q_j generate $\mathbb{Z}/p_j^{n_j}\mathbb{Z}$, and $0 \le m_j \le p_j^{n_j} - 1$. Then $2P = O$ if and only if $m_j = 0$ when p_j is odd and $m_j = 0$ or $2^{n_j - 1}$ otherwise. This means that the order $|\Gamma[2]|$ of the group

$$\Gamma[2] = \{P \in \Gamma \,|\, 2P = O\}$$

is equal to 2^q and we obtain

$$[\Gamma: 2\Gamma] = 2^r |\Gamma[2]|. \tag{7.6}$$

Clearly

$$|\Gamma[2]| = \begin{cases} 4 & \text{if } -A \in \mathbb{Q}^{\times 2}; \\ 2 & \text{otherwise.} \end{cases} \tag{7.7}$$

Now the composition of the homomorphisms $\phi: E \to \bar{E}$ and $\bar{\phi}: \bar{E} \to \bar{\bar{E}} \cong E$ is multiplication by ± 2. Identifying the isomorphic curves E and $\bar{\bar{E}}$ and writing $\bar{\Gamma}$ for $\bar{E}(\mathbb{Q})$, we have

$$\Gamma \supseteq \bar{\phi}(\bar{\Gamma}) \supseteq 2\Gamma = \bar{\phi}\phi(\Gamma).$$

Therefore,

$$[\Gamma: 2\Gamma] = [\Gamma: \bar{\phi}(\bar{\Gamma})][\bar{\phi}(\bar{\Gamma}): 2\Gamma]. \tag{7.8}$$

Applying Theorem 2.35, with $G = \bar{\Gamma}$, $H = \phi(\Gamma)$, and $f = \bar{\phi}$, we get

$$[\bar{\phi}(\bar{\Gamma}): \bar{\phi}\phi(\Gamma)] = \frac{[\bar{\Gamma}: \phi(\Gamma)]}{[\operatorname{Ker} \bar{\phi}: \operatorname{Ker} \bar{\phi} \cap \phi(\Gamma)]}. \tag{7.9}$$

But by Theorems 7.1 and 7.2, $\operatorname{Ker}(\bar{\phi}) = \{O, \mathbf{0}\}$ and $\mathbf{0} \in \phi(\Gamma)$ if and only if $-A \in \mathbb{Q}^{\times 2}$. Hence

$$[\operatorname{Ker}(\bar{\phi}): \operatorname{Ker}(\phi) \cap \phi(\Gamma)] = \begin{cases} 1 & \text{if } -A \in \mathbb{Q}^{\times 2}; \\ 2 & \text{otherwise.} \end{cases} \tag{7.10}$$

Therefore by equations (7.6)–(7.10),

$$\begin{aligned} 2^r &= \frac{[\Gamma: 2\Gamma]}{|\Gamma[2]|} \\ &= \frac{[\Gamma: \bar{\phi}(\bar{\Gamma})][\bar{\Gamma}: \phi(\Gamma)]}{|\Gamma[2]|[\operatorname{Ker}(\bar{\phi}): \operatorname{Ker}(\bar{\phi}) \cap \phi(\Gamma)]} \\ &= \frac{[\Gamma: \bar{\phi}(\bar{\Gamma})][\bar{\Gamma}: \phi(\Gamma)]}{4} \end{aligned}$$

and we have a formula for the rank r:

$$2^r = \frac{[\Gamma: \bar{\phi}(\bar{\Gamma})][\bar{\Gamma}: \phi(\Gamma)]}{4}. \tag{7.11}$$

As before let $\beta\colon \mathbb{Q}^\times \to \mathbb{Q}^\times/\mathbb{Q}^{\times 2}$ denote the homomorphism assigning to each x in $\mathbb{Q}^\times$ its coset, sometimes also denoted by x. We define a map $\alpha\colon \Gamma \to \mathbb{Q}^\times/\mathbb{Q}^{\times 2}$ by

$$\alpha(P) = \begin{cases} 1 & \text{if } P = O, \\ \beta(A), & \text{if } P = \mathbf{0}, \\ \beta(x(P)) & \text{if } x(P) \neq 0. \end{cases} \tag{7.12}$$

THEOREM 7.4. *The map α is a homomorphism from Γ into $\mathbb{Q}^\times/\mathbb{Q}^{\times 2}$ with* $\mathrm{Ker}(\alpha) = \bar{\phi}(\bar{\Gamma})$. *Consequently,*

$$\Gamma/\bar{\phi}(\bar{\Gamma}) \cong \alpha(\Gamma).$$

PROOF. To prove that α is a homomorphism, it is enough to prove that if $P_1 + P_2 + P_3 = O$, then

$$\prod_{j=1}^{3} \alpha(P_j) = 1,$$

because then

$$\alpha(P_1 + P_2) = \alpha(-P_3) = \alpha(P_3)$$
$$= \frac{1}{\alpha(P_3)} = \alpha(P_1)\alpha(P_2).$$

Suppose $P_j (j = 1, 2, 3)$ lie on the line $y = mx + b$ and E. Then $x_j = x(P_j)$ are the three solutions of the equation

$$(mx + b)^2 = x^3 + Ax$$

or

$$x^3 - m^2x^2 + (A - 2bm)x - b^2 = 0. \tag{7.13}$$

If $x_1x_2x_3 \neq 0$, then $x_1x_2x_3 = b^2$, so $\prod_{j=1}^{3} \alpha(P_j) = \prod_{j=1}^{3} \beta(x_j) = \beta(x_1x_2x_3) = \beta(b^2) = 1$. If $x_1x_2x_3 = 0$, then one of x_j say, $x_3 = 0$. So $b = 0$ and $P_3 = \mathbf{0}$. From (7.13), the product of the other two roots $x_1x_2 = A$. Therefore $\alpha(P_1 + P_2) = \alpha(-\mathbf{0}) = \alpha(\mathbf{0}) = \beta(A) = \beta(x_1x_2) = \beta(x_1)\beta(x_2) = \alpha(P_1)\alpha(P_2)$. The fact that $\mathrm{Ker}(\alpha) = \bar{\phi}(\bar{\Gamma})$ follows at once from Theorem 7.2. □

COROLLARY 7.5. *The formula* (7.11) *can be written as*

$$2^r = \frac{|\alpha(\Gamma)|\,|\bar{\alpha}(\bar{\Gamma})|}{4}, \tag{7.14}$$

where $\bar{\alpha}\colon \bar{\Gamma} \to \mathbb{Q}^\times/\mathbb{Q}^{\times 2}$ is the homomorphism defined in the same way as α.

7.4. Computation of $\alpha(\Gamma)$

The subgroup $\alpha(\Gamma)$ of $\mathbb{Q}^\times/\mathbb{Q}^{\times 2}$ consists of those $\beta(x)$ such that $x = x(P)$ for some P in Γ. For $P = (x, y)$ in Γ, we have seen that

$$x = \frac{s}{T^2}, \qquad y = \frac{u}{T^3},$$

with $(s, T) = (u, T) = 1$, $T \geq 1$.

Case 1. If $u \neq 0$, then $s \neq 0$. Put P in (7.1) to get

$$u^2 = s^3 + AsT^4.$$

If $(A, s) = d$, then $A = dA_1$, $s = ds_1$ and therefore, $u = du_1$ and the above equation becomes

$$u_1^2 = s_1(ds_1^2 + A_1T^4)$$

Now s_1 and $ds_1^2 + A_1T^4$ are coprime, so choosing the sign of d properly, we see that

$$s_1 = S^2 \quad \text{and} \quad ds_1^2 + A_1T^4 = U^2.$$

Hence

$$dS^4 + \frac{A}{d}T^4 = U^2. \tag{7.15}$$

with $S, T \geq 1$ and $(A/d, S) = 1$. Also we have $x = dS^2/T^2$ and $\alpha(P) = \beta(d)$.

Case 2. If $u = 0$, then either $P = \mathbf{0}$ in which case $\alpha(P) = \alpha(\mathbf{0}) = \beta(A) \in \alpha(\Gamma)$ or $x^2 + A = 0$. So $\alpha(\Gamma)$ always contains $\beta(A)$ and $\pm\beta(x) \in \alpha(\Gamma)$ if and only if $-A = x^2 (x \in \mathbb{N})$.

Thus, we see that if $z \in \alpha(\Gamma)$, then $z = 1$, $\beta(A)$, $\pm\beta(x)$ (if $-A = x^2$, $x \in \mathbb{N}$) or $z = \beta(d)$ for a divisor d of A, such that (7.15) has an integer solution with

$$(A/d, S) = 1 \quad \text{and} \quad S, T \geq 1. \tag{7.16}$$

Conversely, for a divisor d of A, any integer solution of (7.15) satisfying (7.16) leads to a point $P = (x, y)$ in Γ, such that $\beta(x) = \beta(d)$. To see this multiply (7.15) throughout by d^2S^2/T^6 and put $x = dS^2/T^2$, $y = dUS/T^3$.

Summarizing, we have the following theorem.

THEOREM 7.6. *The group* $\alpha(\Gamma)$ *consists of* 1, $\beta(A)$, $\pm\beta(x)$ (*if* $-A = x^2$, $x \in \mathbb{N}$), *and those* $\beta(d)$ *such that* d *is a* (*positive or negative*) *divisor of* A *with the property that* (7.15) *has an integer solution satisfying* (7.16).

A similar statement holds for $\bar{\alpha}(\bar{\Gamma})$.

7.5. Examples

EXAMPLE 7.7. Let E be defined by

$$y^2 = x^3 - x.$$

Here $A = -1$. The possibilities for d are $d = \pm 1$. Since $\alpha(\Gamma)$ already contains 1 and $\beta(A) = -1$, it is not necessary to check for the solvability of (7.15). So $\alpha(\Gamma) = \{1, -1\}$. For $\bar{A} = 4$, $\bar{d} = \pm 1, \pm 2, \pm 4$. If $\bar{d} < 0$, then $\bar{A}_1 = \bar{A}/\bar{d} < 0$ and (7.15) can have no solution satisfying (7.16). Also $\beta(4) = 1$. So the only case we must consider is $\bar{d} = 2$. The equation (7.15) must have a solution for $\bar{d} = 2$, because $|\alpha(\Gamma)|\,|\bar{\alpha}(\bar{\Gamma})|$ is at least 4. In fact, the solution is $S = T = 1$ and $U = 2$. So $\bar{\alpha}(\bar{\Gamma}) = \{1, 2\}$. Now by (7.14), the rank $r = 0$.

EXAMPLE 7.8. Let E be defined by

$$y^2 = x^3 - 5x.$$

$A = -5$, so $\alpha(\Gamma)$ contains $1, -5$. The other possibilities for d to be considered are $d = -1$ and 5. The equation (7.15) is symmetric in d and A/d and it is enough to check its solvability for either of d or A/d. For $d = 5$, (7.15) has a solution $S = T = 1$, $U = 2$, and so $\alpha(\Gamma) = \{\pm 1, \pm 5\}$.

$\bar{A} = 20$ and $\bar{d} = \pm 1, \pm 2, \pm 4, \pm 5, \pm 10, \pm 20$. For $\bar{d} < 0$, (7.15) can have no nontrivial solution. Also $\beta(\bar{A}) = \beta(20) = 5$ and $\beta(4) = 1$. Because $5 \in \bar{\alpha}(\bar{\Gamma})$ and $\bar{\alpha}(\bar{\Gamma})$ is a group, $10 \in \bar{\alpha}(\bar{\Gamma}) \Leftrightarrow 2 \in \bar{\alpha}(\bar{\Gamma})$. So we have to check (7.15) for solvability only for $\bar{d} = 2$. Suppose

$$2S^4 + 10T^4 = U^2$$

has a solution satisfying (7.16). Obviously $U \neq 0$, otherwise $-5 \in \mathbb{Q}^{\times 2}$. If $5 \mid U$, then $5 \mid 2S^4$, so A_1 and S are not coprime, contradicting (7.16). Reducing the above equation modulo 5, we have

$$2S^4 = U^2.$$

In $\mathbb{F}_5^\times$, $S^4 = 1$, so $2 = U^2$. But we have seen that 2 is not a square in $\mathbb{F}_5^\times$ (Examples 2.51). Hence (7.15) has no solution satisfying (7.16), so neither 2 nor 10 is in $\bar{\alpha}(\bar{\Gamma})$ and $\bar{\alpha}(\bar{\Gamma}) = \{1, 5\}$. By (7.14) now the rank $r = 1$.

EXAMPLE 7.9. Let E be

$$y^2 = x^3 - 17x.$$

$A = -17$ and $1, -17 \in \alpha(\Gamma)$. We need to check (7.15) only for one of the other divisors $d = -1, 17$. Let $d = -1$. Then

$$-S^4 + 17T^4 = U^2$$

has a solution $S = T = 1$, $U = 4$ and $-1 \in \alpha(\Gamma)$, which also implies that $17 \in \alpha(\Gamma)$. Hence $\alpha(\Gamma) = \{\pm 1, \pm 7\}$. $\bar{A} = 4 \cdot 17$ and $\bar{\alpha}(\bar{\Gamma})$ contain $1, 17$. $\bar{d} = \pm 1, \pm 2, \pm 4, \pm 17, \pm 2 \cdot 17, \pm 4 \cdot 17$. Because $\bar{A} > 0$, $\bar{d}$ cannot be negative. Also $\beta(4) = 1$, $\beta(4 \cdot 17) = 17$. So we discard all $\bar{d}$ except $\bar{d} = 2$. Now $S = T = 1$ and $U = 6$ is a solution of

$$2S^4 + 34T^4 = U^2.$$

Hence $\bar{\alpha}(\Gamma) = \{1, 2, 17, 2 \cdot 17\}$ and $2^r = (4 \cdot 4)/4 = 4$, which implies that the rank $r = 2$.

EXERCISE 7.10. Show that $r_{\mathbb{Q}}(E) = 3$, for $E\colon y^2 = x^3 - 82x$.

7.6. Points of Finite Order

We now turn to the computation of $E(\mathbb{Q})_{\text{tor}}$. Unlike the rank, thanks to a result obtained independently by Nagell and Lutz [2], $E(\mathbb{Q})_{\text{tor}}$ can be determined completely in a finite number of steps.

As pointed out earlier, if E is defined by $y^2 = x^3 + Ax + B$ with A, B in $\mathbb{Q}$, then for any c in $\mathbb{Q}^\times$ the curve E is isomorphic to

$$y^2 = x^3 + c^4 AX + c^6 B.$$

Because the torsion subgroups of isomorphic groups are isomorphic, we will again, without loss of generality, assume that E is defined by an equation with coefficients in $\mathbb{Z}$. We again follow Tate [4]. The fundamental result on points of finite order of $E(\mathbb{Q})$ is the following theorem.

THEOREM 7.11 (*Lutz-Nagell*). *Suppose E is defined by*

$$y^2 = x^3 + Ax + B \qquad (A, B \in \mathbb{Z}) \tag{7.17}$$

with the discriminant $\Delta = -4A^3 - 27B^2 \neq 0$. *If* $P = (x, y)$ *in* $E(\mathbb{Q})$ *is a point of finite order, then* $x, y \in \mathbb{Z}$. *Moreover, either* $y = 0$ *or* y^2 *divides* Δ.

Before giving the proof, we introduce an important concept: the valuation.

Suppose $A = \mathbb{Z}$ or $k[x]$, where k is any field. Let K be the quotient field of A, i.e.,

$$K = \begin{cases} \mathbb{Q} & \text{if } A = \mathbb{Z}, \\ k(x) & \text{if } A = k[x]. \end{cases}$$

We call an irreducible monic polynomial in $k[x]$ a prime. Let p be a fixed prime of A. Any α in $K^\times$ can be uniquely written as

$$\alpha = p^r \cdot a/b, \qquad a, b \in A, b \neq 0,$$

where $r \in \mathbb{Z}$ and p does not divide ab. We define a function

$$v_p: \qquad K \to \mathbb{Z} \cup \{\infty\}$$

called the *valuation of K at p*, by $v_p(\alpha) = r$. We put $v_p(0) = \infty$. Clearly, $v_p(\alpha)$ is the exponent to which p appears in the factorization of α into powers (positive, negative, or zero) of primes and has the following properties:

$$\begin{aligned} &1.\ v_p(\alpha\beta) = v_p(\alpha) + v_p(\beta),\\ &2.\ v_p(\alpha + \beta) \geq \min(v_p(\alpha), v_p(\beta)),\\ &3.\ \text{if } v_p(\alpha) \neq v_p(\beta), \text{ then}\\ &\quad v_p(\alpha + \beta) = \min(v_p(\alpha), v_p(\beta)). \end{aligned} \tag{7.18}$$

It is easy to check [use (7.18)] that for a fixed prime p of A,

$$A_{(p)} = \{\alpha \in K \mid v_p(\alpha) \geq 0\}\}$$

is a subring of K, called the *localization of A at p*. Moreover, $A_{(p)} \supseteq A$ and A is the intersection of $A_{(p)}$ for all primes p. Suppose r is a positive integer and $\alpha, \beta \in K$. We write

$$\alpha \equiv \beta (\operatorname{mod} p^r) \tag{7.19}$$

if $v_p(\alpha - \beta) \geq r$. Clearly (7.19) is an equivalence relation.

Proof of Theorem 7.11. If $P = (x, y)$ is a point of finite order of $E(\mathbb{Q})$, it is enough to prove that x, y are in every localization $\mathbb{Z}_{(p)}$ of $\mathbb{Z}$. So let p be any but fixed prime.

We have seen that $P = (s/T^2, u/T^3)$ with $(s, T) = (u, T) = 1$. Therefore, if $v_p(x) < 0$, then $v_p(x) = -2r$, $v_p(y) = -3r$ with $r \geq 1$. For $r \geq 1$, we put

$$\begin{aligned} E_r &= \{(x, y) \in E(\mathbb{Q}) \mid v_p(x) \leq -2r\}\\ &= \{(x, y) \in E(\mathbb{Q}) \mid v_p(y) \leq -3r\}. \end{aligned}$$

Obviously,

$$E(\mathbb{Q}) \supseteq E_1 \supseteq E_2 \supseteq \cdots$$

We will show that if $P \in E(\mathbb{Q})_{\text{tor}}$, then P is not in E_1. Under the rational substitution

$$x = \frac{t}{s}, \qquad y = \frac{1}{s},$$

i.e., (7.20)

$$s = \frac{1}{y}, \qquad t = \frac{x}{y}$$

(7.17) becomes

$$E^*: s = t^3 + Ats^2 + Bs^3. \tag{7.21}$$

The zero of $E(\mathbb{Q})$ is taken by (7.20) to $(0, 0)$ and it sets up a one-to-one correspondence between $\{P \in E \mid P \neq O \text{ and } y(P) \neq 0\}$ and $\{(s, t) \in E^* \mid s \neq 0\}$. Moreover, it maps a straight line $y = mx + b$ to $s = -(m/b)t + 1/b$ (if $b \neq 0$) or $t = 1/m$ (if $b = 0$). Thus collinear points on E are taken to collinear points on E^*. We use $(0, 0)$ as the zero element for the group structure on E^* for which the map $\phi: E \to E^*$ defined by

$$\phi(P) = \begin{cases} \left(\dfrac{1}{y}, \dfrac{x}{y}\right) & \text{if } P = (x, y) \text{ with } y \neq 0, \\ (0, 0) & \text{if } P = 0 \text{ or } y(P) = 0 \end{cases}$$

is an isomorphism between the groups $E^*(\mathbb{Q})$ and $E(\mathbb{Q})/\{P \in E(\mathbb{Q}) \mid 2P = O\}$. From (7.20) it is obvious that

$$\begin{aligned} \phi(E_r) = E_r^* &= \{(s, t) \in E^*(\mathbb{Q}) \mid v_p(s) \geq 3r\} \\ &= \{(s, t) \in E^*(\mathbb{Q}) \mid v_p(t) \geq r\} \end{aligned}$$

and to prove that no point of finite order of $E(\mathbb{Q})$ is in E_1, it suffices to show that if $P \in E^*(\mathbb{Q})_{\text{tor}}$, then P is not in E_1^*. First we show that for each $r \geq 1$, E_r^* is a subgroup of $E^*(\mathbb{Q})$.

If $P_i = (s_i, t_i)$, $i = 1, 2$, then to get $P_1 + P_2$ we intersect the line

$$s = \mu t + \lambda \tag{7.22}$$

joining P_1 and P_2 with (7.21). If (s_3, t_3) is the third point of intersection, it is clear from (7.20) that $P_1 + P_2 = (-s_3, -t_3)$. If $t_1 = t_2$, but $s_1 \neq s_2$, then $t_3 = t_1 = t_2$ and there is nothing to prove. So assume that this is not the case.

Because P_1 and P_2 lie on the curve (7.21), we obtain

$$\begin{aligned} s_1 - s_2 = (t_1 - t_2)(t_1^2 + t_1 t_2 + t_2^2) &+ B(s_1 - s_2)(s_1^2 + s_1 s_2 + s_2^2) \\ &+ A(t_1 s_1^2 - t_2 s_2^2). \end{aligned}$$

If we write $t_1 s_1^2 - t_2 s_2^2 = (t_1 - t_2)s_2^2 + t_1(s_1^2 - s_2^2)$, the above equation can be written as

$$\begin{aligned} (s_1 - s_2)[1 - B(s_1^2 + s_1 s_2 + s_2^2) - At_1(s_1 + s_2)] \\ = (t_1 - t_2)[t_1^2 + t_1 t_2 + t_2^2 + As_2^2]. \end{aligned}$$

Therefore, the slope μ of (7.22) is given by

$$\mu = \begin{cases} \dfrac{s_1 - s_2}{t_1 - t_2} = \dfrac{t_1^2 + t_1 t_2 + t_2^2 + As_2^2}{1 - B(s_1^2 + s_1 s_2 + s_2^2) - At_1(s_1 + s_2)} & \text{if } t_1 \neq t_2 \\ \left.\dfrac{ds}{dt}\right|_{P_1} = \dfrac{As_1^2 + 3t_1^2}{1 - 2As_1 t_1 - 3Bs_1^2} & \text{if } P_1 = P_2. \end{cases} \tag{7.23}$$

Because t_1, t_2, t_3 are the three roots of

$$\mu t + \lambda = t^3 + A(\mu t + \lambda)^2 t + B(\mu t + \lambda)^3,$$

i.e.,

$$t^3 + \frac{2A\lambda\mu + 3B\lambda\mu^2}{1 + A\mu^2 + B\mu^3} t^2 + \cdots = 0$$

we get

$$t_1 + t_2 + t_3 = -\frac{2A\lambda\mu + 3B\lambda\mu^2}{1 + A\mu^2 + B\mu^3}. \tag{7.24}$$

If $P_j = (s_j, t_j) \in E_r^*$ with $v_p(t_j) \geq r$ $(j = 1, 2)$, using (7.18) it follows from (7.23) that $v_p(\mu) \geq 2r$ and then from (7.22), $v_p(\lambda) \geq 3r$. Consequently by (7.24), $v_p(t_1 + t_2 + t_3) \geq 5r$. If $t(p)$ denotes the t-coordinate of a point P on E^*, this shows that

$$t(P_1 + P_2) \equiv t(P_1) + t(P_2) (\bmod p^{5r}). \tag{7.25}$$

So E_r^* is a subgroup of $E^*(\mathbb{Q})$.

Now we show that if $P \in E^*(\mathbb{Q})_{\text{tor}}$, then P is not E_1^*. Suppose $P \in E_1^*$ and $m = \text{ord}(P)$. There are two cases:

(1) *p does not divide m.* Choose $r \geq 1$ such that P is in E_r^* but not in E_{r+1}^*. Then by (7.25),

$$mt(P) \equiv t(mP) \equiv 0 (\bmod p^{2r}).$$

But m and p are coprime, so $v_p t(P) \geq 2r$, implying that $P \in E_{2r}^*$. Because $2r \geq r + 1$, this is a contradiction.

(2) *p divides m.* Put $m = pm_1$ and $P' = m_1 P$. Then $P' \in E_1^*$ and $p = \text{ord}(P')$. Choose $r \geq 1$ such that $P' \in E_r^* - E_{r+1}^*$. Then

$$0 = t(pP') \equiv p \cdot t(P') (\bmod p^{3r}).$$

This implies that $v_p(t(P')) \geq 3r - 1$, so that $P' \in E_{3r-1}^*$. But for $r \geq 1$, $3r - 1 > r + 1$. Hence $P' \in E_{r+1}^*$, again a contradiction.

Finally, we show that if $P = (x, y) \in E(\mathbb{Q})_{\text{tor}}$ and $y \neq 0$, then $y^2 | \Delta = \Delta(f)$, where $f(x) = x^3 + Ax + B$. From the duplication formula, we obtain

$$x(2P) = \mu^2 - 2x \qquad \text{with } \mu = \frac{f'(x)}{2y}.$$

Because x and $x(2P) \in \mathbb{Z}$, $\mu \in \mathbb{Z}$, which implies that $y | f'(x) = 3x^2 + A$. It is easy to check (by division algorithm) that in the ring $\mathbb{Z}[x]$, $(3x^2 + A)^2(3x^2 + 4A) - (4A^3 + 27B^2) \equiv 0 (\bmod x^3 + Ax + B)$, which shows that $y^2 | - 4A^3 - 27B^2 = \Delta$. □

7.7. Examples

On any elliptic curve E defined by (7.17) there are only finitely many rational points of finite order. We put

$$E(\mathbb{Z}, \Delta) = \{(x, y) \in E(\mathbb{Q}) | x, y \in \mathbb{Z} \text{ and either } y = 0 \text{ or if } y \neq 0, \text{ then } y^2 | \Delta\} \cup \{O\}.$$

It is clear that

$$\{O\} \subseteq E(\mathbb{Q})_{\text{tor}} \subseteq E(\mathbb{Z}, \Delta).$$

There may be points in $E(\mathbb{Z}, \Delta)$ that are not of finite order. [For example, if E is defined by

$$y^2 = x^3 - x + 1,$$

the point $(1, 1) \in E(\mathbb{Z}, \Delta)$. But we have seen [Example 5.26(1)] that $6P = (1/4, 7/8)$. So $6P$ and hence P is a point of infinite order.] We can discard those points of $E(\mathbb{Z}, \Delta)$ that are of infinite order in a finite number (depending only on A and B) of steps by finding a multiple mP of P such that either

1. $y(mP)$ is not an integer

or

2. $y^2(mP)$ does not divide Δ.

This completely determines $E(\mathbb{Q})_{\text{tor}}$ for any elliptic curve E.

One would expect $E(\mathbb{Q})_{\text{tor}}$ to get larger as the number $\tau(\Delta)$ of (positive) divisors of Δ gets larger, but this is not the case. In fact, no matter what elliptic curve E we take, the order of $E(\mathbb{Q})_{\text{tor}}$ never exceeds 16. This follows from a deep theorem of Mazur. We shall not prove this theorem (for proof, see Ref. 3).

THEOREM 7.12* (*Mazur*). *For any elliptic curve E defined over* $\mathbb{Q}$, $E(\mathbb{Q})_{\text{tor}}$ *is isomorphic to one of the following* 15 *groups* [*all these groups occur as* $E(\mathbb{Q})_{\text{tor}}$]:

$$\mathbb{Z}/m\mathbb{Z}, \qquad 1 \le m \le 10 \quad \text{or} \quad m = 12$$

$$\mathbb{Z}/2\mathbb{Z} \times \mathbb{Z}/2m\mathbb{Z}, \qquad 1 \le m \le 4.$$

Now we give examples of some of these 15 groups.

EXAMPLE 7.13. E is defined by

$$y^2 = x^3 - x + 1.$$

The discriminant $\Delta = -23$. The only possibility for y^2 is 1 and $E(\mathbb{Z}, \Delta) = \{O, (\pm 1, \pm 1), (0, \pm 1)\}$. If $P = (1, 1)$, $2P = (-1, 1)$, $3P = (0, -1)$, $4P = (3, -5)$ and $y(4P)$ does not divide Δ. If $Q = (1, -1)$, then $Q = -P$ and we see that O is the only point of finite order and $E(\mathbb{Q})_{\text{tor}}$ is trivial.

EXAMPLE 7.14. E is defined as follows:

$$y^2 = x^3 - 1.$$

The discriminant $\Delta = -3^3$. If $y = 0$, then $x = 1$. The only possibilities for nonzero y^2 are 1 and 3^2, but then there is no integer x satisfying this equation. Obviously, $(1, 0)$ is a point of order 2 and $E(\mathbb{Q})_{\text{tor}} \cong \mathbb{Z}/2\mathbb{Z}$.

EXAMPLE 7.15. E is the Fermat curve

$$y^2 = x^3 - 432.$$

We have seen [Example 5.26(2)] that $E(\mathbb{Q})_{\text{tor}} \cong \mathbb{Z}/3\mathbb{Z}$.

EXAMPLE 7.16. E is given by

$$y^2 = x^3 - 2x + 1$$

has discriminant $\Delta = 5$. If $y = 0$, then $x = 1$, so $(1, 0)$ is a point of order 2. If $y \neq 0$, then $y^2 = 1$, so $x = 0$. If $P = (0, 1)$, it is easily seen that $2P = (1, 0)$ and thus P is a point of order 4. Since $(0, -1) = -\mathrm{P}$, we have $(0, -1) = 3P$ and $E(\mathbb{Q})_{\text{tor}}$ is a cyclic group of order 4 generated by $(0, 1)$, i.e., $E(\mathbb{Q})_{\text{tor}} \cong \mathbb{Z}/4\mathbb{Z}$.

EXAMPLE 7.17. If E is

$$y^2 = x^3 - x,$$

each of $P_1 = (1, 0)$, $P_2 = (0, 0)$, $P_3 = (-1, 0)$ is a point of order 2 and for i, j, k (distinct), $P_i + P_j = P_k$. There is no other integer point (x, y) with y^2 dividing $\Delta = 4$. Hence

$$E(\mathbb{Q})_{\text{tor}} \cong \mathbb{Z}/2\mathbb{Z} \times \mathbb{Z}/2\mathbb{Z}.$$

EXAMPLES 7.18. The curve E defined by

$$y^2 = x^3 + 1$$

has discriminant $\Delta = -3^3$.

$$E(\mathbb{Z}, \Delta) = \{O, (-1, 0), (0, \pm 1), (2, \pm 3)\}.$$

All these points are of finite order and hence $E(\mathbb{Q})_{\text{tor}} = E(\mathbb{Z}, \Delta)$. One can check that $E(\mathbb{Q})_{\text{tor}}$ is a cyclic group of order 6 generated by $P = (2, 3)$.

EXAMPLE 7.19. Consider the elliptic curve

$$E\colon y^2 = x^3 - 43x + 166.$$

We have seen [Exercise 5.27(2)] that $P = (3, 8)$ is a point of order 7. By Mazur's theorem, $E(\mathbb{Q})_{\text{tor}}$ is a cyclic group of order 7 generated by $(3, 8)$.

As an exercise, complete the list in Mazur's theorem.

7.8. Application to Congruent Numbers

We can call a positive rational A a congruent number if it is the area of a right triangle with all sides rational. This definition may seem more general than the one given earlier (Definition 1.33), but we may find a positive integer c, such that c^2A is an integer. Moreover, c^2A is a congruent number if and only if A is a congruent number. Thus it is sufficient to study: *when is a square-free integer $A > 0$ a congruent number*? In this section we discuss an interesting connection between the property of a square-free positive integer A being a congruent number and the rank of the elliptic curve

$$y^2 = x^3 - A^2x. \tag{7.26}$$

We shall make extensive use of the properties (7.18) of the valuation map

$$v_p\colon \mathbb{Q}^\times \to \mathbb{Z}$$

in the proof of the next two theorems. When we talk about, say, the numerator of a rational number x, we shall always assume that x is in the lowest form, i.e., $x = m/n$ with $(m, n) = 1$, $n \geq 1$.

LEMMA 7.20. *Suppose A is a positive square-free integer and E is the elliptic curve defined by* (7.26). *Let $P = (x, y) \in E(\mathbb{Q})$ with $y \neq 0$. Then the numerator of $x(2P)$ is coprime to A and $y(2P) \neq 0$.*

PROOF. It is easy to check that $x(2P)$ is a square. In fact,

$$x(2P) = \left(\frac{x^2 + A^2}{2y}\right)^2.$$

In order to prove the lemma we show that for all prime divisors $p \neq 2$ of A, $v_p(x^2 + A^2) \leq v_p(y)$ and $v_2(x^2 + A^2) \leq v_2(2y)$. There are two cases to consider:

Case 1. First suppose that p is any prime, not necessarily a prime divisor of A, and $v_p(x) \neq v_p(A)$. If $v_p(x) < v_p(A)$, then $v_p(x^2 + A^2) = 2v_p(x)$. But by (7.26), which can also be written as

$$y^2 = x(x + A)(x - A), \tag{7.27}$$

$v_p(x) = \frac{2}{3}v_p(y)$. So

$$v_p(x^2 + A^2) = 2v_p(x) = \tfrac{4}{3}v_p(y) \leq v_p(y).$$

The last inequality is true because, A being square-free, we have $v_p(A) \leq 1$, so $v_p(x)$ and $v_p(y)$ are both ≤ 0.

If $v_p(x) > v_p(A)$, then $v_p(x^2 + A^2) = 2v_p(A) = 0$ or 2, according as $v_p(A) = 0$ or 1, and in any case, by (7.26),

$$\begin{aligned} v_p(y) &= \tfrac{1}{2}[v_p(x) + 2v_p(A)] \\ &\geq 2v_p(A) = v_p(x^2 + A^2). \end{aligned}$$

Case 2. Now let $p \mid A$ and $v_p(x) = v_p(A) = 1$. By (7.26) again, $2v_p(y) \geq 3$. But $2v_p(y)$ is even. So $v_p(y) \geq 2$. We have nothing to prove unless

$v_p(x^2 + A^2) > 2$. We will show that this cannot happen. Suppose it does. First note that by (7.27), either

$$v_p(x + A) \geq 2 \quad \text{or} \quad v_p(x - A) \geq 2,$$

so that it is clear from

$$x^2 + A^2 = (x \pm A)^2 \mp 2xA \tag{7.28}$$

that $p^3 | 2xA$. If p is odd, this implies that p^2 divides either x or A, a contradiction. The prime $p = 2$ can appear in the numerator of $(x^2 + A^2)/2y$ only if $v_p(x^2 + A^2) \geq 4$. But then by (7.28), $v_p(xA) \geq 3$. But this is a contradiction, because $v_p(xA) = v_p(x) + v_p(A) = 2$.

To prove that $y(2P) \neq 0$, let us put $mP = (x_m, y_m)$. Note that since $y = y_1 \neq 0$, we have $x = x_1 \neq 0$. We have seen that $x_2 \in \mathbb{Q}^{\times 2}$. So if $y_2 = 0$, then $x_2 = \pm A$. This is not possible, because A is square free. □

LEMMA 7.21. *Suppose E is defined by* (7.26) *and $P = (X, Y) \in E(\mathbb{Q})$, such that the numerator of X is coprime to A and $Y \neq 0$. Then*

1. *the numerator of $x(2P)$ is coprime to A: and*
2. *the denominator of $x(2P)$ is even.*

PROOF. By Lemma 7.20, the numerator of

$$x(2P) = \left(\frac{X^2 + A^2}{2Y}\right)^2$$

is coprime to A. All we have to show is that $v_2(X^2 + A^2) \leq v_2(Y)$.

If A is even, i.e., $2|A$, then $v_2(X) \leq 0$ and from

$$Y^2 = X(X + A)(X - A),$$

$$v_2(Y) = \tfrac{3}{2}v_2(X) = \tfrac{3}{4}v_2(X^2 + A^2) \geq v_2(X^2 + A^2).$$

If A is odd and $v_2(X) \neq v_2(A)$, by case (1) of the proof of Lemma 7.20, $v_2(X^2 + A^2) \leq v_2(Y)$. If A is odd and $v_2(X) = v_2(A) = 0$, there is nothing to prove unless $v_2(X^2 + A^2) \geq 1$. Now by the identity

$$X^2 + A^2 = (X \pm A)^2 \mp 2XA,$$

$v_2(X \pm A) \geq 1$, which shows that $v_2(Y) \geq 1$. To complete the proof, we show that $v_2(X^2 + A^2) = 1$. Suppose $v_2(X^2 + A^2) \geq 2$. By the above identity, either $v_2(X) > 0$ or $v_2(A) > 0$, a contradiction. □

THEOREM 7.22. *Suppose the elliptic curve E defined by* (7.26) *has a rational point P with nonzero y-coordinate. Then for some $m = 1, 2$, or 4,*

1. *the numerator of $x(mP)$ is coprime to A;*
2. *the denominator of $x(mP)$ is even; and*
3. $x(mP) \in \mathbb{Q}^{\times 2}$.

PROOF. Apply Lemmas 7.20 and 7.21 successively and note that $x(2P)$ is always square.

COROLLARY 7.23. *If E is given by* (7.26), *then*

$$\begin{aligned} E(\mathbb{Q})_{\text{tor}} &= \{O, \mathbf{0}, (\pm A, 0)\} \\ &\cong \mathbb{Z}/2\mathbb{Z} \times \mathbb{Z}/2\mathbb{Z}. \end{aligned}$$

PROOF. If $P = (x, y) \in E(\mathbb{Q})$ with $y \neq 0$, then for an $m = 1, 2$, or 4, $x(mP)$ is not an integer. By the Nagell–Lutz theorem, P is a point of infinite order. So the order of P is finite if and only if either $P = O$ or $y(P) = 0$.

THEOREM 7.24. *Suppose A is a square-free positive integer and E is the elliptic curve defined by* (7.26). *The equations*

$$\begin{aligned} x^2 + Ay^2 &= z^2, \\ x^2 - Ay^2 &= t^2 \end{aligned} \tag{7.29}$$

have a nontrivial solution (*a solution with* $y \neq 0$) *in* $\mathbb{Q}$ *if and only if* $r_{\mathbb{Q}}(E) > 0$.

PROOF. First suppose that (7.29) has a nontrivial solution. We may suppose that all $x, y, z, t \in \mathbb{N}$. As we saw in Chapter 1, we may further assume that x, y, z, t are coprime in pairs. One can check that $y \neq 1$.

Multiplying the two equations in (7.29), we obtain

$$\left(\frac{ztx}{y^3}\right)^2 = \left(\frac{x^2}{y^2}\right)^3 - A^2\frac{x^2}{y^2}.$$

Hence $P = (X, Y)$ with noninteger coordinates

$$X = \frac{x^2}{y^2}, \qquad Y = \frac{ztx}{y^3}$$

is a point of infinite order on (7.26) and so $r_{\mathbb{Q}}(E) > 0$. Conversely, let $E(\mathbb{Q})$ have a point $P = (X, Y)$ of infinite order. Then $Y \neq 0$ and

$$X = \frac{s}{t^2}, \qquad Y = \frac{u}{t^3} \tag{7.30}$$

with

$$(s, t) = (u, t) = 1 \quad \text{and} \quad t \geq 1. \tag{7.31}$$

By Theorem 7.22, we may assume that

$$s \geq 1 \quad \text{is odd,} \quad t \geq 2 \text{ is even,} \quad \text{and} \quad (s, A) = 1. \tag{7.32}$$

Since (X, Y) is on (7.26), from (7.30) we obtain

$$u^2 = s(s + At^2)(s - At^2). \tag{7.33}$$

Using (7.31) and (7.32), it is easily checked that the three factors on the right-hand side of (7.33) are positive, mutually coprime and therefore each factor is a square. If $s = v^2$, we have

$$v^2 + At^2 = m^2,$$

$$v^2 - At^2 = n^2.$$

This completes the proof. □

REMARK 7.25. We have seen that 1 is not a congruent number (Theorem 1.35). This now follows also from Theorem 7.24 and Example 7.7.

References

1. B. J. Birch and H. P. F. Swinnerton-Dyer, Notes on elliptic curves II, *J. Reine Angew. Math.* **218**, 79–108 (1965).
2. E. Lutz, Sur l'équation $y^2 = x^3 - Ax - B$ dans les corps p-adiques, *J. Reine Angew. Math.* **177**, 237–247 (1937).
3. B. Mazur, Rational points on modular curves, *Modular Functions of One Variable V, Lecture Notes in Mathematics* Vol. 601, Springer Verlag, Berlin (1977).
4. J. T. Tate, Rational points on elliptic curves, Phillips Lectures given at Haverford College, 1961 (unpublished).

8

Equations Over Finite Fields

8.1. Riemann Hypothesis

We have seen that for each prime p, there is a field $\mathbb{F}_p$ of p elements. In fact, given any prime p and an integer $r \geq 1$, there is one and only one field $\mathbb{F}_q$ of $q = p^r$ elements. The field $\mathbb{F}_q \supseteq \mathbb{F}_p$ and for each α in $\mathbb{F}_q$, $p\alpha = 0$. Conversely, any finite field is $\mathbb{F}_q$ for some $q = p^r$ (cf. Ref. 18). The field $\mathbb{F}_q$ is characterized by the property

$$f(X) = X^q - X = \prod_{\alpha \in \mathbb{F}_q} (X - \alpha).$$

If x, y are in a field K containing $\mathbb{F}_q$, then

1. $(x + y)^q = x^q + y^q$;
2. $(xy)^q = x^q y^q$ and $\alpha^q = \alpha$ if $\alpha \in \mathbb{F}_q$.

The second assertion needs no proof. The first follows from the fact that for $j = 1, \ldots, q - 1$, the binomial coefficients in

$$(x + y)^q = \sum_{j=0}^{q} \binom{q}{j} x^j y^{q-j}$$

are multiples of q, so that only the first and the last terms survive. Thus the function ϕ: $K \to K$ defined by $\phi(x) = x^q$ is an endomorphism of the ring K. It is called the *Frobenius endomorphism.*

We are interested in counting the number N_q of solutions in $\mathbb{F}_q \times \mathbb{F}_q$ of the equation

$$Y^2 = f(X), \tag{8.1}$$

where $f(X) = AX^3 + BX^2 + CX + D \in \mathbb{F}_q[X]$, $A \neq 0$ and $f(X)$ has no repeated root. We suppose that $p \neq 2, 3$, so that (8.1) can be written (see the proof of Corollary 5.21) as

$$Y^2 = X^3 + aX + b \qquad (a, b \in \mathbb{F}_q). \tag{8.2}$$

Together with the hypothetical point O at infinity, these solutions form an abelian group of order $N'_q = N_q + 1$. This is the group of $\mathbb{F}_q$-rational points on the elliptic curve E defined by (8.2). Let $q = p$. In 1924, Artin conjectured the following estimate for $N_p : |N_p - p| \le 2\sqrt{p}$. Actually, an equivalent form of this inequality is the analog for the field of rational functions on the curve (8.2) of what Riemann conjectured much earlier for the field of rational numbers, and is well known as the Reimann hypothesis. (For the equivalence, see Ref. 19.) Gauss was the first to study the behavior of N_p, as p varies, for the curve

$$Y^2 = X^3 - 432. \tag{8.3}$$

[If $p \neq 2, 3$, (8.3) is the same as $X^3 + Y^3 = 1$. See example 5.6.] In fact, he gave a precise formula for N_p.

THEOREM 8.1* (*Gauss 1801*). *Let N_p be the number of solutions in $\mathbb{F}_p \times \mathbb{F}_p$ of $Y^2 = X^3 - 432$, $p \neq 2, 3$. Then*

1. $N_p = p$ *for* $p \equiv 2 \pmod 3$;
2. *If* $p \equiv 1 \pmod 3$, *there are integers A, B, unique up to sign, such that $4p = A^2 + 27B^2$. If the sign of A is so chosen that $A \equiv 1 \pmod 3$, then $N_p = p + A - 2$. In particular, $|N_p - p| \le 2\sqrt{p}$.*

For its proof see Chap. 8 of Ref. 13.

Artin's conjecture was proved by Hasse in 1936. Later (1948) Weil generalized it to his famous theorem (the Riemann hypothesis for curves over finite fields) and made some intriguing conjectures which are known as Weil conjectures.

THEOREM 8.2* RIEMANN HYPOTHESIS FOR CURVES OVER FINITE FIELDS (*Weil*). *The number N_q of points with coordinates in $\mathbb{F}_q$ on an irreducible, non-singular curve defined over $\mathbb{F}_q$ and of genus g satisfies*

$$|N_q - q| \le 2g\sqrt{q}. \tag{8.4}$$

Manin [15] gave a completely elementary proof of Hasse's theorem (see also Chap. 10 of Ref. 11) and a valuation theoretic proof is due to Zimmer [26]. Weil's proof of Riemann hypothesis depends heavily on algebraic geometry. For further comments on Weil's proof, see Ref. 7, pp. 208–211. A somewhat simpler proof was given by Roquette [16]. An elementary proof was later initiated by Stepanov [22] and completed by Schmidt (cf. Ref. 18). A very elegant but less elementary proof based on Stepanov's method is by Bombieri [4]. For a self-contained account of Bombieri's proof, see Ref. 19. We shall give Manin's proof of Hasse's theorem.

THEOREM 8.3 (*Hasse*). *The number N_p of solutions in $\mathbb{F}_p \times \mathbb{F}_p$ of the equation*

$$Y^2 = X^3 + aX + b(a, b \in \mathbb{F}_p) \tag{8.5}$$

with $\Delta = -4a^3 - 27b^2$ in $\mathbb{F}_p^\times$ satisfies the inequality

$$|N_p - p| \leq 2\sqrt{p}. \tag{8.6}$$

REMARK 8.4. When the curve is projective, there is an extra point (the point at infinity), so that the total number of points now is $N'_q = N_q + 1$ and (8.4) becomes

$$|N'_q - (q+1)| \leq 2g\sqrt{q}.$$

8.2. Manin's Proof of Hasse's Theorem

First let us remark that in Theorem 8.3, p can be replaced by $q = p^r$ and the proof is still valid without any change.

Suppose E_1, E_2 are two elliptic curves defined over a field K. Then E_1 is a *twist* of E_2 over K if E_1, E_2 become isomorphic over a finite extension L of K. Let $K = \mathbb{F}_p(x)$, the function field in one variable over $\mathbb{F}_p$, and suppose E is the elliptic curve defined by (8.5). If E_λ is the elliptic curve defined by

$$\lambda Y^2 = X^3 + aX + b, \tag{8.7}$$

where

$$\lambda = \lambda(x) = x^3 + ax + b, \tag{8.8}$$

then E and E_λ are both defined over K and E_λ is a twist of E over K.

To prove Theorem 8.3, we shall consider the group $E_\lambda(K)$ of K-rational points on E_λ. Carrying out the necessary modification of formulas (5.28)–(5.33), it is easily seen that if (X, Y), $(X_j, Y_j) \in E_\lambda(E)$, $j = 1, 2$ with $(X, Y) = (X_1, Y_1) + (X_2, Y_2,)$, then

$$X = \lambda\left(\frac{Y_1 - Y_2}{X_1 - X_2}\right)^2 - (X_1 + X_2). \tag{8.9}$$

If $(X, Y) = 2(X_1, Y_1)$, then

$$X = \frac{(3X_1^2 + a)^2}{4(X_1^3 + aX_1 + b)\lambda} - 2X_1. \tag{8.10}$$

In the field $K = \mathbb{F}_p(x)$, the equation (8.7) has two obvious solutions: $(x, 1)$, $(x, -1) = -(x, 1)$. A less obvious solution is

$$X_0 = x^p, \qquad Y_0 = (x^3 + ax + b)^{(p-1)/2}.$$

Let

$$(X_n, Y_n) = (X_0, Y_0) + n(x, 1), \qquad n \in \mathbb{Z}.$$

If $(X_n, Y_n) \neq O$, we will show that $X_n \neq 0$. Writing X_n in the lowest form as $X_n = P_n/Q_n$ with Q_n and monic P_n in $\mathbb{F}_p[x]$, we get a well-defined function

$$d : \mathbb{Z} \to \{0, 1, 2, 3, \ldots\}$$

given by

$$d(n) = d_n = \begin{cases} 0, & \text{if } (X_n, Y_n) = O; \\ \deg(P_n) & \text{otherwise.} \end{cases}$$

Manin's proof of Hasse's theorem is based on the following basic identity:

Basic Identity: $d_{n-1} + d_{n+1} = 2d_n + 2.$

The connection between Hasse's theorem and the function $d(n)$ is the following identity:

$$d_{-1} - d_0 - 1 = N_p - p. \tag{8.11}$$

To prove identity (8.11), we need to put the rational function X_{-1} in the lowest form. By the addition formula (8.9), we have

$$X_{-1} = \frac{(x^3 + ax + b)[(x^3 + ax + b)^{(p-1)/2} + 1]^2}{(x^p - x)^2} - (x^p + x)$$
$$= \frac{x^{2p+1} + R(x)}{(x^p - x)^2},$$

where $R(x)$ is a polynomial of degree at most $2p$. To put X_{-1} in the lowest form P_{-1}/Q_{-1} first note that

$$(x^p - x) = x(x-1)\cdots(x-p+1).$$

The only factors to cancel from the denominator are either $(x-r)^2$ with the Legendre symbol

$$\left(\frac{r^3 + ar + b}{p}\right) = (r^3 + ar + b)^{(p-1)/2} = -1$$

or $x - r$ with $r^3 + ar + b = 0$ $(0 \leq r < p)$. If m is the number of factors of the first kind and n the number of factors of the second kind, then

$$d_{-1} = \deg P_{-1} = 2p + 1 - 2m - n.$$

But $d_0 = p$, so that

$$d_{-1} - d_0 = p + 1 - 2m - n. \tag{8.12}$$

Each r in $\mathbb{F}_p$ with $r^3 + ar + b \neq 0$ and

$$\left(\frac{r^3 + ar + b}{p}\right) = 1$$

will give rise to two solutions of (8.5), whereas we get only one solution from $r^3 + ar + b = 0$. Hence

$$N_p = 2(p - m) - n$$

and (8.11) follows from (8.12).

LEMMA 8.5. *The function $d(n)$ is a quadratic polynomial in n. In fact,*

$$d_n = n^2 - (d_{-1} - d_0 - 1)n + d_0.$$

PROOF. The lemma is obviously true for $n = -1, 0$. Suppose it is true for $n - 1$ and n $(n \geq 0)$. By the basic identity,

$$\begin{aligned} d_{n+1} &= 2d_n - d_{n-1} + 2 \\ &= 2[n^2 - (d_{-1} - d_0 - 1)n + d_0] \\ &\quad - [(n-1)^2 - (d_{-1} - d_0 - 1)(n-1) + d_0] + 2 \\ &= (n+1)^2 - (d_{-1} - d_0 - 1)(n+1) + d_0, \end{aligned}$$

proving the lemma for $n + 1$. By induction the lemma holds for all $n \geq -1$. Similarly, it holds for all $n \leq 0$. □

PROOF OF THEOREM 8.3. The quadratic polynomial

$$\begin{aligned} d(x) &= x^2 - (d_{-1} - d_0 - 1)x + d_0 \\ &= x^2 - (N_p - p)x + d_0 \end{aligned}$$

assumes non-negative values for all $n \in \mathbb{Z}$. Hence its discriminant

$$D = (N_p - p)^2 - 4p$$

cannot be positive, because otherwise $d(x)$ will have two real roots α, β such that for some n,

$$n \leq \alpha < \beta \leq n + 1.$$

Moreover, both equalities cannot hold simultaneously because by definition, $d(n)$ cannot be zero for two successive integers. But this is a contradiction because $(\alpha - \beta)^2 = D \in \mathbb{Z}$. Thus $(N_p - p)^2 - 4p \leq 0$, which proves the estimate (8.6). □

8.3. Proof of the Basic Identity

For the proof we need the following lemma.

LEMMA 8.6. *If* $(X_n, Y_n) \neq O$, *then* $\deg P_n > \deg Q_n$. *In particular,* $X_n \neq 0$.

PROOF. To prove that the degree of numerator of a rational function $R(x)$ in $\mathbb{F}_p(x)$ is larger than that of its denominator, we formally evaluate $R(x)$ at $x = \infty$ and show that $R(x)_{|\infty} = \infty$.

The lemma is obviously true for $n = 0$ and for those $n > 0$ for which $(X_{n-1}, Y_{n-1}) = O$. Suppose the lemma is true for a particular $n \geq 0$ for which $(X_{n-1}, Y_{n-1}) \neq O$. The proof for $n \geq 0$ will follow by induction if we show the lemma to hold for $n + 1$ also. Since

$$Y_{n+1}^2 = \frac{X_{n+1}^3 + aX_{n+1} + b}{x^3 + ax + b}, \tag{8.13}$$

it is clear that $\deg P_{n+1} > \deg Q_{n+1}$ if and only if $Y_{n+1}|_\infty \neq 0$.

Suppose $\deg P_{n+1} \leq \deg Q_{n+1}$, i.e., $Y_{n+1}|_\infty = 0$. Because

$$(X_{n+1}, Y_{n+1}) = (X_n, Y_n) + (x, 1),$$

we have

$$Y_{n+1} = \frac{1 - Y_n}{x - X_n}(x - X_{n+1}) - 1,$$

so that

$$0 = Y_{n+1}|_\infty = \left\{\frac{1 - Y_n}{1 - X_n/x}(1 - X_{n+1}/x) - 1\right\}\bigg|_\infty.$$

But $X_{n+1}/x|_\infty = 0$, therefore

$$\frac{1 - Y_n}{1 - X_n/x}\bigg|_\infty = 1. \tag{8.14}$$

From the addition formula (8.9), i.e.,

$$X_{n+1} = \left(\frac{1 - Y_n}{x - X_n}\right)^2 (x^3 + ax + b) - x - X_n,$$

we get

$$\frac{X_{n+1}}{x} = \left(\frac{1 - Y_n}{1 - X_n/x}\right)^2 \left(1 + \frac{a}{x^2} + \frac{b}{x^3}\right) - 1 - \frac{X_n}{x}.$$

Hence by (8.14) and the induction hypothesis, we get

$$0 = \frac{X_{n+1}}{x}\bigg|_\infty = \left\{\left(\frac{1 - Y_n}{1 - X_n/x}\right)^2 \left(1 + \frac{a}{x^2} + \frac{b}{x^3}\right) - 1 - \frac{X_n}{x}\right\}\bigg|_\infty = -\frac{X_n}{x}\bigg|_\infty \neq 0.$$

This contradiction proves the lemma for $n \geq 0$. The induction for $n \leq 0$ is carried out similarly. □

We now prove the basic identity. When one of (X_{n-1}, Y_{n-1}), (X_n, Y_n), (X_{n+1}, Y_{n+1}) is O, the basic identity is a triviality. In fact, if $(X_n, Y_n) = O$, then $X_{n-1} = X_{n+1} = x$ and $d_n = 0$, $d_{n-1} = d_{n+1} = 1$ and there is nothing to prove. If $(X_{n-1}, Y_{n-1}) = O$, then $(X_n, Y_n) = (x, 1)$ and by the addition formula (8.10),

$$X_{n+1} = \frac{x^4 - 2ax^2 - 8bx + a^2}{4(x^3 + ax + b)}.$$

Clearly $d_{n-1} = 0$, $d_n = 1$. It can be checked that the above expression for X_{n+1} is in the lowest form, hence $d_{n+1} = 4$ and the lemma follows. The third possibility $(X_{n+1}, Y_{n+1}) = O$ can be dealt with in a similar way.

Thus we may suppose that none of (X_{n-1}, Y_{n-1}), (X_n, Y_n), (X_{n+1}, Y_{n+1}) is O. In the addition formula (8.9), we bring all the terms to a common denominator and obtain

$$\begin{aligned} X_{n-1} &= \frac{-(xQ_n + P_n)(xQ_n - P_n)^2 + (1 + Y_n)^2(x^3 + ax + b)Q_n^3}{Q_n(xQ_n - P_n)^2} \qquad (8.15) \\ &= \frac{(xQ_n + P_n)(xP_n + aQ_n) + 2bQ_n^2 + 2Y_n(x^3 + ax + b)Q_n^2}{(xQ_n - P_n)^2}, \end{aligned}$$

i.e.,

$$X_{n-1} = \frac{R}{(xQ_n - P_n)^2}. \qquad (8.16)$$

Similarly,

$$\begin{aligned} X_{n+1} &= \frac{-(xQ_n + P_n)(xQ_n - P_n)^2 + (1 - Y_n)^2(x^3 + ax + b)Q_n^3}{Q_n(xQ_n - P_n)^2} \qquad (8.17) \\ &= \frac{(xQ_n + P_n)(xP_n + aQ_n) + 2bQ_n^2 - 2Y_n(x^3 + ax + b)Q_n^2}{(xQ_n - P_n)^2}, \end{aligned}$$

i.e.,

$$X_{n+1} = \frac{S}{(xQ_n - P_n)^2}. \qquad (8.18)$$

It can be checked that on multiplying the above expressions for X_{n-1} and X_{n+1}, we obtain

$$\frac{P_{n-1}P_{n+1}}{Q_{n-1}Q_{n+1}} = \frac{RS}{(xQ_n - P_n)^4} = \frac{(xP_n - aQ_n)^2 - 4bQ_n(xQ_n + P_n)}{(xQ_n - P_n)^2}. \qquad (8.19)$$

If we show that

$$Q_{n-1}Q_{n+1} = (xQ_n - P_n)^2, \tag{8.20}$$

then

$$P_{n-1}P_{n+1} = (xP_n - aQ_n)^2 - 4bQ_n(xQ_n + P_n)$$

and by Lemma 8.6, we obtain

$$\begin{aligned} d_{n-1} + d_{n+1} &= \deg(P_{n-1}P_{n+1}) \\ &= \deg(x^2P_n^2) = 2d_n + 2. \end{aligned}$$

Recall that for an irreducible polynomial $l = l(x)$ in $\mathbb{F}_p[x]$, the valuation map

$$v_l : \mathbb{F}_p(x) \to \mathbb{Z} \cup \{\infty\}$$

is defined to be the power to which l appears in the factorization of rational functions into powers of irreducible polynomials. To prove (8.20), it is enough to show that for each irreducible l,

$$v_l(Q_{n-1}Q_{n+1}) = v_l((xQ_n - P_n)^2). \tag{8.21}$$

It is clear from (8.13) that the rational function $Y_n(x^3 + ax + b)Q_n^2$ is actually a polynomial, hence $R, S \in \mathbb{F}_p[x]$. Moreover, for each n, Q_n is the denominator of X_n in the lowest form. Hence it follows from (8.16) and (8.18) that Q_{n-1} and Q_{n+1} both divide $(xQ_n - P_n)^2$, so that $v_l(xQ_n - P_n) = 0$ implies that $v_l(Q_{n-1}Q_{n+1}) = 0$. Thus all we need to show is that (8.21) holds for those l for which $v_l(xQ_n - P_n) > 0$.

Any such l divides RS. First suppose that l divides only one of R, S; say l divides R but not S. By (8.16), (8.18), and (8.19), $v_l(Q_{n-1}) = 0$ and $v_l(Q_{n+1}) = v_l((xQ_n - P_n)^2)$, which proves (8.20).

Finally, we suppose that a prime divisor l of $xQ_n - P_n$ divides both R and S, so that l divides both $Q_n^3(1 + Y_n)^2(x^3 + ax + b)$ and $Q_n^3(1 - Y_n)^2(x^3 + ax + b)$. Since P_n, Q_n are coprime and l divides $xQ_n - P_n$, l cannot divide Q_n. Hence l divides

$$(1 + Y_n)(x^3 + ax + b) \text{ and } (1 - Y_n)\ (x^3 + ax + b)$$

from which it follows that l divides $x^3 + ax + b$. [From l dividing a rational function $\mu(x)$ we mean that $v_l(\mu(x)) \geq 1$.] The polynomial $x^3 + ax + b$ has no repeated root, so $v_l(x^3 + ax + b) = 1$.

If both $v_l(1 \pm Y_n) \leq 0$, then it follows from

$$v_l(Q_n^3(1 \pm Y_n)(x^3 + ax + b)) = 1$$

and the properties (7.18) of the valuation function that $v_l(R) = v_l(S) = 1$. Since $(xQ_n - P_n)^2$ divides RS,

$$v_l((xQ_n - P_n)^2) = v_l(RS) = 2.$$

If one of $v_l(1 \pm Y_n)$, say $v_l(1 + Y_n) > 0$, then $v_l(1 - Y_n) \leq 0$. In fact, $v_l(1 - Y_n) = 0$, because otherwise $(1 - Y_n)^2\,(x^3 + ax + b)Q_n^3$ cannot be a polynomial. Therefore $v_l(S) = 1$. Suppose $v_l(1 + Y_n) = f$ and $v_l(xQ_n - P_n) = g$. Then

$$2f = v_l((1 + Y_n)^2(1 - Y_n)^2) = v_l((1 - Y_n^2)^2).$$

But

$$\begin{aligned}(1 - Y_n^2)^2 &= \frac{(x^3 + ax - X_n^3 - aX_n)^2}{(x^3 + ax + b)^2}\\ &= \frac{(x - X_n)^2(x^2 + xX_n + X_n^2 + a)^2}{(x^3 + ax + b)^2}.\end{aligned} \tag{8.22}$$

Since l is irreducible, it is obvious (from Theorem 4.11) that the remainders of polynomials in $\mathbb{F}_p[x]$ on division by l form a field. Since l divides both $xQ_n - P_n$ and $x^3 + ax + b$, in this field $x = P_n/Q_n = X_n$ and

$$x^2 + xX_n + X_n^2 + a = 3x^2 + a \neq 0.$$

Therefore, by (8.22),

$$2f = v_l(x - X_n)^2 - 2 = 2g - 2,$$

so that

$$\begin{aligned}v_l((1 + Y_n)^2(x^3 + ax + b)) &= 2f + 1 = 2g - 1\\ &< 2g = v_l((xQ_n - P_n)^2).\end{aligned}$$

This shows that $v_l(R) = 2g - 1$ and $v_l(RS) = 2g$. This completes the proof of the basic identity.

8.4. Analytic Methods

Suppose E is an elliptic curve given in the Weierstrass form

$$y^2 = x^3 + Ax + B \qquad (A, B \in \mathbb{Z}). \tag{8.23}$$

Its discriminant $\Delta = -4A^3 - 27B^2 \neq 0$, hence Δ is divisible only by finitely many primes (called the *bad* primes). If $p > 2$ is a *good* prime, i.e., a prime not dividing Δ, the reduction E_p of (8.23) modulo p is an elliptic curve defined over $\mathbb{F}_p$. We know, in general, very little about the group $E(\mathbb{Q})$ of rational points on E, which is a global object. The next best thing to do is to look at the corresponding local objects, namely the groups $E_p(\mathbb{F}_p)$, and try somehow to use them in obtaining information about $E(\mathbb{Q})$. This is done via Hasse-Weil L function.

Let V be a variety, say a projective variety defined over $\mathbb{Q}$. It is the set of common solutions in a projective space of a finite number of homogeneous polynomial equations

$$f_1(\mathbf{x}) = \cdots = f_m(\mathbf{x}) = 0, \tag{8.24}$$

with coefficients in $\mathbb{Q}$ which in fact, may be assumed to be in $\mathbb{Z}$. We shall assume that V is irreducible and denote by V_p the variety defined over $\mathbb{F}_p$ by the reduction

$$f_{1,p}(\mathbf{x}) = \cdots = f_{m,p}(\mathbf{x}) = 0 \tag{8.25}$$

of (8.24) modulo p. Let $N(r)$ denote the number of solutions of (8.25) in $\mathbb{P}^n(\mathbb{F}_{p^r})$. The *congruence zeta function* $Z(V_p, T)$ of V_p is defined by

$$Z(V_p, T) = \exp\left[\sum_{r=1}^{\infty} \frac{N(r)}{r} T^r\right],$$

where for $z \in \mathbb{C}$,

$$\exp(z) = \sum_{j=0}^{\infty} \frac{z^j}{j!}.$$

Conversely, if we know $Z(V_p, T)$, we can obtain $N(r)$ by the formula

$$N(r) = \frac{1}{(r-1)!} \frac{d^r}{dT^r} \log Z(V_p, T)_{|T=0}.$$

EXAMPLE 8.7. The variety $V = \mathbb{P}^n$ is defined by the zero polynomial. For each $r \geq 1$, the set $\mathbb{P}^n(\mathbb{F}_{p^r})$ consists of nonzero vectors $(x_1, \ldots, x_{n+1})$, $x_i \in \mathbb{F}_{p^r}$, with two vectors identified if one is a multiple of the other by a scalar in $\mathbb{F}_{p^r}^{\times}$. Thus

$$N(r) = \frac{p^{r(n+1)} - 1}{p^r - 1} = \sum_{j=0}^{n} p^{rj}$$

and

$$\log Z(V_p, T) = \sum_{j=0}^{n} \sum_{r=1}^{\infty} \frac{p^{rj}}{r} T^r$$

$$= -\sum_{j=0}^{n} \log(1 - p^j T),$$

which shows that

$$Z(V_p, T) = \frac{1}{(1-T)(1-pT)\cdots(1-p^nT)}.$$

Thus $Z(V_p, T)$ is a rational function of T. This is a special case of the famous conjectures made in 1949 by A. Weil [25]. For details see Ref. 8.

WEIL CONJECTURES. *Suppose V_p is "smooth". Then*

1. *$Z(V_p, T)$ is a rational function of T;*
2. *$Z(V_p, T)$ has a functional equation;*
3. *The Riemann hypothesis holds for $Z(V_p, T)$.*

For elliptic curves and abelian varieties these were proved by Weil himself. For an elementary proofs see Refs. 18 and 19. In the general case, the rationality was established by Dwork [9] and the proof of the Riemann hypothesis was given by P. Deligne (1973). For a survey of Deligne's proof, see Ref. 8, pp. 147–160.

In the case of elliptic curves, it can be shown (cf. Refs. 19 or 21) that

$$Z(E_p, T) = \frac{1 - a_p T + pT^2}{(1-T)(1-pT)},$$

where

$$a_p = a_p(E) = p - N_p.$$

By estimate (8.6), the discriminant of the polynomial $1 - a_p T + pT^2$ is nonpositive. Thus the two roots of this polynomial are complex conjugates of each other and if α, $\bar{\alpha}$ are the reciprocals of these roots, we have

$$1 - a_p T + pT^2 = (1 - \alpha T)(1 - \bar{\alpha} T),$$

with

$$\alpha + \bar{\alpha} = a_p, \qquad |\alpha| = |\bar{\alpha}| = \sqrt{p}. \tag{8.26}$$

We make a change of variable $T = p^{-s}$ ($s = \sigma + it$) and define the *local zeta function* of *E at p* by

$$\zeta(E_p, s) = Z(E_p, p^{-s}) = \frac{1 - a_p p^{-s} + p^{1-2s}}{(1-p^{-s})(1-p^{1-s})}. \tag{8.27}$$

Note that by (8.26), if $\zeta(E_p, s) = 0$, then $|p^s| = \sqrt{p}$, i.e., $\sigma = 1/2$. This is the traditional form of Riemann hypothesis.

The local zeta function $\zeta(E_p, s)$ is defined for those primes $p > 2$ that do not divide the discriminant $\Delta = -4A^3 - 27B^2$ of E. If $p|2\Delta$, we put

$$\zeta(E_p, s) = \frac{1}{(1-p^{-s})(1-p^{1-s})}$$

and define the *global zeta function* of E by

$$\zeta(E, s) = \prod_p \zeta(E_p, s).$$

The *Hasse-Weil L function*

$$L(E, s) = \prod_{p \nmid 2\Delta} (1 - a_p p^{-s} + p^{1-2s})^{-1} \tag{8.28}$$

of E is related to $\zeta(E, s)$ and *the Reimann zeta function*

$$\zeta(s) = \prod_p (1 - p^{-s})^{-1}$$

by

$$\zeta(E, s) = \zeta(s)\zeta(s-1)/L(E, s).$$

Using Hasse's theorem, it easily follows that the *Euler product* in (8.28) for $L(E, s)$ converges for $\sigma > 3/2$.

We now return to the group $E(\mathbb{Q})$ of rational points on the elliptic curve (8.23). In Chapter 7, we computed the rank $r_{\mathbb{Q}}(E)$ of some elliptic curve which turned out to be rather small. In general, it is extremely difficult to find curves of high rank. All the ranks that have been calculated so far turned out to be very small. In fact, in most cases $r_{\mathbb{Q}}(E) = 0$, 1, or 2 (cf. Table 1 in Ref. 3). Mestre (1985) found a curve of rank fourteen, the largest known rank. One may ask:

Are there curves of arbitrarily large rank?

It is still an open question and the answer is predicted by the following conjecture.

CONJECTURE 8.8. (*Cassels* [5]-*Tate* [23]). *There are elliptic curves defined over* $\mathbb{Q}$ *with arbitrarily large rank* $r_{\mathbb{Q}}(E)$.

Another open problem is the computation of the rank $r_{\mathbb{Q}}(E)$. Conjecturally, it is related to the numbers N_p as follows:

We have remarked that $L(E, s)$ converges for $\sigma > 3/2$, where $s = \sigma + it$.

CONJECTURE 8.9. (*Hasse-Weil-Deuring*). *The function* $L(E, s)$ *can be extended analytically to the whole plane.*

CONJECTURE 8.10 (*Birch-Swinnerton-Dyer*). *If the analytic function* $L(E, s)$ *is expanded at* $s = 1$ [*the center of the functional equation relating* $L(E, s)$ *to* $L(E, 2-s)$] *as*

$$L(E, s) = a_g(s-1)^g + a_{g+1}(s-1)^{g+1} + \cdots$$

with $g \geq 0$ *and* $a_g \neq 0$, *then* $g = r_{\mathbb{Q}}(E)$.

The *weak form* of this conjecture is

$$r_{\mathbb{Q}}(E) > 0 \text{ if and only if } g > 0.$$

Hasse was the first to make Conjecture 8.9. Weil proved it in some special cases. Deuring proved it for elliptic curves with complex multiplication. (For the definition of complex multiplication see Definition A.24.) This includes the cases $y^2 = x^3 + Ax$ and $y^2 = x^3 + B$.

Some partial results have been obtained on Conjecture 8.10, e.g., the following theorem.

THEOREM 8.11* (*Coates–Wiles* [6]). *Suppose E is an elliptic curve defined by* (8.23) *and has complex multiplication. If $r_{\mathbb{Q}}(E) > 0$, then $g > 0$.*

For results in the other direction [i.e., $g > 0$ implies $r_{\mathbb{Q}}(E) > 0$] see Refs. 12 and 17.

8.5. Application to the Congruent Number Problem

In this section we show how the following well-known conjecture follows from the conjecture of Birch and Swinnerton-Dyer.

CONJECTURE 8.12. *Suppose $A > 0$ is a square-free integer. If $A \equiv 5, 6, 7$ (mod. 8), then A is a congruent number.*

Throughout this section E will be the elliptic curve

$$y^2 = x^3 - A^2x \tag{8.29}$$

associated to the congruent number problem. We know that A is a congruent number if and only if $r_{\mathbb{Q}}(E) > 0$ (Theorems 1.34 and 7.24) and assuming Birch and Swinnerton-Dyer conjecture, this is equivalent to the vanishing of $L(E, s)$ at $s = 1$. [$L(E, s)$ has analytic continuation to the whole plane by Deuring's result.] For this elliptic curve the functional equation for $L(E, s)$ is given by the following theorem (cf. Chap. II in Ref. 14).

THEOREM 8.13*. *For E as in* (8.29) *define its conductor N by*

$$N = \begin{cases} 32A^2 & \text{if } A \text{ is odd;} \\ 16A^2 & \text{if } A \text{ is even} \end{cases}$$

and put

$$\Phi(s) = \left(\frac{\sqrt{N}}{2\pi}\right)^s \Gamma(s) L(E, s).$$

Then we have the following functional equation:

$$\Phi(s) = w\Phi(2 - s), \tag{8.30}$$

where the root number

$$w = \begin{cases} 1 & \text{if } A \equiv 1, 2, 3 \pmod 8, \\ -1 & \text{if } A \equiv 5, 6, 7 \pmod 8). \end{cases}$$

Here $\Gamma(s)$ is the *Euler's gamma function*, a generalization of the factorial: $n! = 1, 2, 3 \cdots n$. It is defined for all complex numbers s with $\mathrm{Re}(s) > 0$ by

$$\Gamma(s) = \int_0^\infty t^{s-1} e^{-t}\, dt.$$

For $n \in \mathbb{Z}$, $n \geq 0$, it can be easily checked that

$$n! = \Gamma(n+1).$$

We now show how Conjecture 8.12 follows from this functional equation and the weak Birch and Swinnerton-Dyer conjecture. By the hypothesis on A, $w = -1$. Put $s = 1$ in (8.30) to get $\Phi(1) = -\Phi(1)$, i.e., $\Phi(1) = 0$, so that $L(E, 1) = 0$. The Birch and Swinnerton-Dyer conjecture now implies that $r_{\mathbb{Q}}(E) > 0$ and hence A is a congruent number.

Recently Tunnell [24] showed how one can decide in a finite number of steps that a given square-free integer A is *not* a congruent number. If one assumes the weak Birch and Swinnerton-Dyer conjecture for the curves (8.29), Tunnell's theorem completely solves the problem of classifying all the congruent numbers.

THEOREM 8.14* (*Tunnell*). *For a square-free integer $A > 0$, let $n_1(A)$ denote the number of triplets (x, y, z) in $\mathbb{Z}^3$ such that*

$$A = 2x^2 + y^2 + 32z^2 \qquad \text{if } A \text{ is odd};$$

$$A/2 = 4x^2 + y^2 + 32z^2 \qquad \text{if } A \text{ is even}.$$

Similarly, let $n_2(A)$ denote the number of $(x, y, z) \in \mathbb{Z}^3$ such that

$$A = 2x^2 + y^2 + 8z^2, \qquad \text{if } A \text{ is odd};$$

$$A/2 = 4x^2 + y^2 + 8z^2, \qquad \text{if } A \text{ is even}.$$

If A is a congruent number, then

$$n_2(A) = 2n_1(A). \tag{8.31}$$

Conversely, the weak Birch and Swinnerton-Dyer conjecture and (8.31) *imply that A is a congruent number.*

For the proof of this theorem and a detailed discussion of the congruent numbers, see Ref. 14.

8.6. Remarks on Curves of Higher Genus

There are only finitely many points of finite order on the elliptic curve (8.23). These points have necessarily integer coordinates. However, as we have seen, an integer point on E need not be of finite order. [(1, 1) is a point of infinite order on $E: y^2 = x^3 - x + 1$.] Thus one might ask the following question:

Is it possible for an elliptic curve to have infinitely many integer points?

A deep theorem of Siegel (cf. Ref. 20) says no.

THEOREM 8.15* (*Siegel*). *Let $f(x, y) \in \mathbb{Z}[x, y]$ be an irreducible polynomial. There are only finitely many integer points on the curve*

$$f(x, y) = 0, \tag{8.32}$$

if its genus $g \geq 1$.

Actually, a conjecture of Mordell predicted that such curves can have infinitely many rational points only if its genus $g = 1$. This has been proved recently by Faltings [10].

THEOREM 8.16* (*Faltings*). *There are only finitely many rational points on* (8.32), *if its genus $g \geq 2$.*

Theorems 8.15 and 8.16 are noneffective, i.e., there is no way to determine in a finite number of steps these finitely many points. However, in the case of genus one, a result of Baker and Coates [2] solves the problem of integer points completely.

THEOREM 8.17* (*Baker–Coates*). *Let $f(x, y)$ be an absolutely irreducible polynomial of degree n with integer coefficients of absolute value at most H. If the genus of*

$$f(x, y) = 0$$

is one, then all the integer points (x, y) on this curve satisfy

$$\max(|x|, |y|) < \exp\exp\exp[(2H)^{10^{n^{10}}}]. \tag{8.33}$$

In the special case we have been dealing with, the bound is not as bad as in (8.33).

THEOREM 8.18* (*Baker* [1]). *If (x, y) is an integer point on an elliptic curve*

$$y^2 = x^3 + Ax + B \qquad (A, B \in \mathbb{Z})$$

and $H = \max(|A|, |B|)$, then

$$\max(|X|, |Y|) < \exp[(10^6 H)^{10^6}].$$

References

1. A. Baker, The diophantine equation $y^2 = ax^3 + bx^2 + cx + d$, *J. London Math. Soc.* **43** 1-9, (1968).
2. A. Baker and J. Coates, Integer points on curves on genus 1, *Proc. Cambridge Phil. Soc.* **67** 595-602, (1970).
3. B. J. Birch and H. P. F. Swinnerton-Dyer, Notes on elliptic curves II, *J. Reine Angew. Math.* **218**, 79-108 (1965).
4. E. Bombieri, Counting points over finite fields (d'après S. A. Stepanov), *Sem. Bourbaki*, Exposé 430 (1972-73).
5. J. W. S. Cassels, Diophantine equations with special reference to elliptic curves, *J. London Math. Soc.* **41**, 193-291 (1966).
6. J. Coates and A. Wiles, On the conjecture of Birch and Swinnerton-Dyer, *Invent. Math.* **39**, 223-251 (1977).
7. G. Cornell and J. H. Silverman (eds.), *Arithmetic Geometry*, Springer Verlag, New York (1986).
8. J. Dieudonné, *History of Algebraic Geometry*, Wadsworth, Belmont, California (1985).
9. B. Dwork, On the rationality of zeta function, *Amer. J. Math.* **82**, 631-648 (1960).
10. G. Faltings, Endlichkeitssätze für abelsche Varietäten über Zahlkörpern, *Invent. Math.* **73**, 349-366 (1938).
11. A. Gelfond and Yu. Linnik, *Elementary Methods in Analytic Number Theory*, Fizmatgiz, Moscow (1962).
12. B. Gross and D. Zagier, Heegner points and derivatives of *L*-series, *Invent. Math.* **84**, 225-320 (1986).
13. K. Ireland and M. Rosen, *A Classical Introduction to Modern Number Theory*, GTM 84, Springer Verlag, New York, (1982).
14. N. Koblitz, *Introduction to Elliptic Curves and Modular Forms*, GTM 97, Springer Verlag, New York (1984).
15. Yu I. Manin, On cubic congruences to a prime modulus, *Izv. Akad. Nauk USSR, Math. Ser.* **20**, 673-678 (1956).
16. P. Roquette, Arithmetischer Beweis der Riemannschen Vermutung in Kongruenzzetafunktionenkörpern Belibingen Geschlechts, *J. Reine Angew. Math.* **191**, 199-252 (1953).
17. K. Rubin, Tate-Shafarevich groups and *L*-functions of elliptic curves with complex multiplication, *Invent. Math.* **89**, 527-560 (1987).
18. W. M. Schmidt, *Lectures on Equations over Finite Fields: An Elementary Approach*, Part I, Lecture Notes in Math; No. 536, Springer Verlag, Berlin (1976).
19. W. M. Schmidt, Lectures on Equations over Finite Fields: An Elementary Approach, Part II (unpublished).
20. C. L. Siegel, Über einige Anwendungen diophantischer Approximationen, *Abh. Preuss. Akad. Wiss. Phys. Math.* **K1** Nr. 1 (1929).
21. J. H. Silverman, *The Arithmetic of Elliptic Curves*, GTM 106, Springer Verlag, New York (1986).
22. S. A. Stepanov, The number of points of a hyperelliptic curve over a prime field, *Izv. Akad. Nauk USSR, Math. Ser.* **33**, 1171-1181 (1969).
23. J. Tate, The arithmetic of elliptic curves, *Invent. Math.* **23**, 179-206 (1974).
24. J. Tunnell, A classical diophantine problem and modular forms of weight 3/2, *Invent Math*, **72**, 323-334 (1983).
25. A. Weil, Number of solutions of equations in finite fields, *Bull. Am. Math. Soc.* **55**, 497-508 (1949).
26. H. G. Zimmer, An elementary proof of the Riemann hypothesis for an elliptic curve over a finite field, *Pacific J. Math.* **36**, 267-278 (1971).

Appendix
Weierstrass Theory

The purpose of this appendix is to show that the group $E(\mathbb{C})$ of $\mathbb{C}$-rational points on an elliptic curve E is isomorphic to the torus $T = \mathbb{C}/L$ for a lattice L.

A.1. Review of Complex Analysis

A function $f: \mathbb{C} \to \mathbb{C}$ is called *meromorphic* if f is analytic (holomorphic) except for poles. Poles are by definition isolated, i.e., for each pole ω of f, there is a positive real number r such that f is analytic in the punctured disk

$$\{z \in \mathbb{C} \,|\, 0 < |z - \omega| < r\}$$

of radius r and centered at ω. It is easy to see that the meromorphic functions form a field under the addition and multiplication of functions. The proof of the following theorem can be found in any book on complex analysis, e.g., Ref. 1.

THEOREM A.1 (*Liouville*). *If $f(z)$ is analytic and bounded on $\mathbb{C}$ then $f(z)$ is a constant function.*

A.2. Elliptic Functions

Suppose $L = \mathbb{Z}\omega_1 \oplus \mathbb{Z}\omega_2 = \{m\omega_1 + n\omega_2 \,|\, m, n \in \mathbb{Z}\}$ is a lattice with periods ω_1, ω_2. We may assume that ω_1/ω_2 is in the upper half plane

$$\{z \in \mathbb{C} \,|\, \mathrm{Im}(z) > 0\}.$$

Then L is a subgroup of the additive group $\mathbb{C}$ of complex numbers and the quotient $T = \mathbb{C}/L$ is a torus [cf. Example 2.20(2)].

DEFINITION A.2. A meromorphic function

$$f: \mathbb{C} \to \mathbb{C}$$

is said to be an *elliptic function relative to* a lattice L (or *doubly periodic* with periods ω_1, ω_2), if

$$f(z + \omega) = f(z) \tag{A.1}$$

for each ω in L.

The condition (A.1) is equivalent to the following two conditions:

1. $f(z + \omega_1) = f(z)$;
2. $f(z + \omega_2) = f(z)$.

Let $\mathbb{E}(L)$ denote the set of elliptic functions relative to L. The set $\mathbb{E}(L)$ is clearly a field under the usual addition and multiplication of functions.

The complex plane is a (disjoint) union

$$\mathbb{C} = \bigcup_{\omega \in L} (\omega + T)$$

of the translates $\omega + T$ of the torus T, with T being identified with the fundamental parallelogram

$$\{x\omega_1 + y\omega_2 | 0 \leq x < 1, 0 \leq y < 1\}.$$

If $z \in \mathbb{C}$, let $z = \omega + \tau$ with $\omega \in L$, $\tau \in T$. Therefore, for any $f(z) \in \mathbb{E}(L)$,

$$f(z) = f(\omega + \tau) = f(\tau),$$

i.e., $f(z)$ is completely determined by its values on T.

THEOREM A.3. *Suppose $f(z) \in \mathbb{E}(L)$. If $f(z)$ has no poles in T, then $f(z)$ is constant.*

PROOF. Since $f(z)$ has no poles in T, it has no poles in $\mathbb{C}$ either. Moreover, $f(z)$ is bounded on $\mathbb{C}$ for if

$$X = \{x\omega_1 + y\omega_2 | 0 \leq x, y \leq 1\},$$

then

$$f(X) = \{f(z) | z \in X\}$$

is bounded (the image of a compact set under a continuous function is compact) and $f(\mathbb{C}) = f(X)$. Since $f(z)$ is analytic and bounded on $\mathbb{C}$, it must be a constant function, by Liouville's theorem. □

THEOREM A.4. *For any lattice*

$$L = \mathbb{Z}\omega_1 \oplus \mathbb{Z}\omega_2$$

the series

$$\sum_{\omega \in L-\{0\}} \frac{1}{\omega^\sigma}$$

(of complex numbers) converges absolutely for all real numbers $\sigma > 2$.

PROOF. Let L_r denote the set of those lattice points of L that lie on the parallelogram

$$P_r = \{x\omega_1 \pm r\omega_2; \pm r\omega_1 + y\omega_2 \,|\, |x|, |y| \leq r\}.$$

If a is the shortest distance of P_1 from the origin O, then each ω in L_r satisfies the inequality

$$|\omega| \geq ra$$

and therefore

$$\frac{1}{|\omega|} \leq \frac{1}{ra}.$$

If we put

$$a_r = \sum_{\omega \in L_r} \frac{1}{\omega^\sigma},$$

then it is clear that (cf. Fig. A.1)

$$\sum_{\omega \in L-\{0\}} \frac{1}{\omega^\sigma} = \sum_{r=1}^{\infty} a_r. \tag{A.2}$$

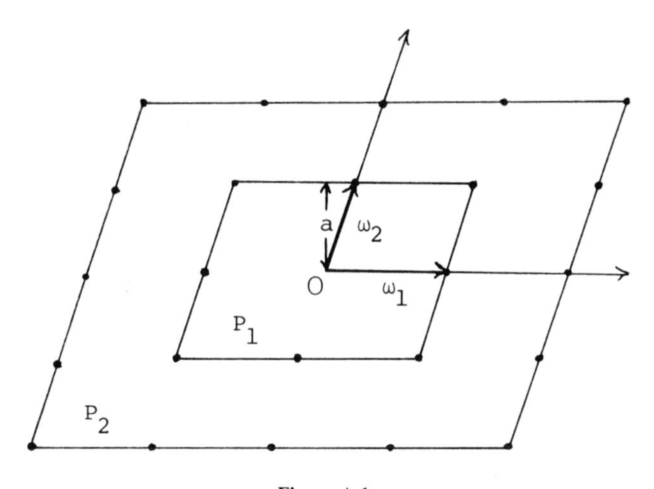

Figure A.1

Each L_r consists of $8r$ points. Therefore

$$|a_r| = \left| \sum_{\omega \in L_r} \frac{1}{\omega^\sigma} \right|$$

$$\leq \sum_{\omega \in L_r} \frac{1}{|\omega|^\sigma}$$

$$\leq \frac{8r}{(ar)^\sigma} = \frac{c}{r^{\sigma-1}},$$

where the constant $c = 8/a^\sigma$. Now the theorem follows by comparing (A.2) with the series

$$c \sum_{r=1}^{\infty} \frac{1}{r^{\sigma-1}}$$

which converges for $\sigma > 2$. □

Definition A.5. When $\sigma > 2$ is an integer the series

$$G_\sigma(L) = \sum_{\omega \in L-\{0\}} \frac{1}{\omega^\sigma}$$

$$= \sum_{\substack{m,n=1 \\ (m,n) \neq (0,0)}}^{\infty} \frac{1}{(m\omega_1 + n\omega_2)^\sigma}$$

is called the *Eisenstein series (for L) of weight* σ.

If σ is odd, the terms cancel in pairs:

$$\frac{1}{\omega^\sigma} + \frac{1}{(-\omega)^\sigma} = 0$$

and $G_\sigma(L) = 0$. Thus we assume that $\sigma = 2k$ is even. The series $G_{2k}(L)$ is convergent for $k > 1$.

Corollary A.6. *Suppose $X \subseteq \mathbb{C} - L$ is a compact set. Then the series*

$$\sum_{\omega \in L-\{0\}} \left[\frac{1}{(z-\omega)^2} - \frac{1}{\omega^2} \right] \tag{A.3}$$

converges absolutely and uniformly on X.

Proof. Since X is bounded, there is a positive integer N (depending only on X) such that for each z in X,

$$|z - \omega| \geq \tfrac{1}{2}|\omega| \tag{A.4}$$

for all the lattice points ω in L satisfying the inequality

$$|\omega| > N.$$

Now

$$\begin{aligned}\frac{1}{(z-\omega)^2} - \frac{1}{\omega^2} &= \frac{2\omega z - z^2}{\omega^2(z-\omega)^2} \\ &= \frac{\omega z + z(\omega - z)}{\omega^2(z-\omega)^2} \\ &= \frac{z}{\omega(z-\omega)^2} + \frac{z}{\omega^2(z-\omega)}.\end{aligned}$$

So by (A.4), for $|\omega| > N$,

$$\begin{aligned}\left|\frac{1}{(z-\omega)^2} - \frac{1}{\omega^2}\right| &\le \left|\frac{z}{\omega(z-\omega)^2}\right| + \left|\frac{z}{\omega^2(z-\omega)}\right| \\ &\le \frac{4|z|}{|\omega|^3} + \frac{2|z|}{|\omega|^3} \\ &= \frac{6|z|}{|\omega|^3}.\end{aligned}$$

If we ignore the sum

$$f(z) = \sum_{|\omega| \le N} \left[\frac{1}{(z-\omega)^2} - \frac{1}{\omega^2}\right]$$

of finitely many terms, which does not affect the convergence, the series (A.3) is easily seen to be convergent by comparing it with

$$6|z| \sum_{\omega \in L - \{0\}} \frac{1}{\omega^3}.$$

The function $|f(z)|$ and $|z|$ are both bounded on the compact set X, say by M. Therefore the error

$$E_N = \sum_{|\omega| > N} \left[\frac{1}{(z-\omega)^2} - \frac{1}{\omega^2}\right]$$

can be bounded independent of z by

$$6M \sum_{|\omega| > N} \frac{1}{|\omega|^3}.$$

Hence the convergence is uniform. □

The *Weierstrass ℘-function* is defined by

$$\wp(z) = \frac{1}{z^2} + \sum_{\omega \in L-\{0\}} \left[\frac{1}{(z-\omega)^2} - \frac{1}{\omega^2}\right].$$

To show its dependence on L or on ω_1, ω_2, we sometimes write it as $\wp(z, L)$ or $\wp(z, \omega_1/\omega_2)$.

THEOREM A.7. *$\wp(z, L) \in \mathbb{E}(L)$. It has a double pole at each $\omega \in L$ and these are its only poles.*

PROOF. For any $\omega \in L$, the function

$$\wp(z) - \frac{1}{(z-\omega)^2}$$

is bounded in any compact neighborhood U of ω that does not intersect L. In U, $1/(z-\omega)^2$ is the principal part of $\wp(z)$, which proves the second statement in the theorem. [It also shows that $\text{Res}_\omega \wp(z) = 0$.]

Because of uniform convergence, we can differentiate $\wp(z)$ termwise to get

$$\wp'(z) = -2 \sum_{\omega \in L} \frac{1}{(z-\omega)^3}.$$

For a fixed ω_0 in L, the sum over ω is the same as the sum over $\omega - \omega_0$, as ω runs over all the lattice points. So

$$\begin{aligned}\wp'(z+\omega_0) &= -2 \sum_{\omega \in L} \frac{1}{[z-(\omega-\omega_0)]^3} \\ &= -2 \sum_{\omega \in L} \frac{1}{(z-\omega)^3} \\ &= \wp'(z),\end{aligned}$$

i.e., $\wp'(z) \in \mathbb{E}(L)$.

To show that $\wp(z) \in \mathbb{E}(L)$, we first note that $\wp(z)$ is an even function, for

$$\begin{aligned}\wp(-z) &= \frac{1}{z^2} + \sum_{\omega \in L-\{0\}} \left[\frac{1}{(z+\omega)^2} - \frac{1}{(-\omega)^2}\right] \\ &= \wp(z),\end{aligned}$$

because summing over ω is the same as summing over $-\omega$, $\omega \in L$.

To show that $\wp(z) \in \mathbb{E}(L)$, we must show that $\wp(z+\omega_j) = \wp(z)$, for $j = 1, 2$. If $j = 1$,

$$\wp(z+\omega_1) - \wp(z) = c, \tag{A.5}$$

a constant, because in view of $\wp'(z)$ being in $\mathbb{E}(L)$, $\wp'(z+\omega_1) - \wp'(z) = 0$.

Because $\wp(z)$ is even, evaluating (A.5) at $z = -\omega_1/2$, we obtain

$$c = \wp\left(\frac{\omega_1}{2}\right) - \wp\left(-\frac{\omega_1}{2}\right) = 0.$$

This proves that $\wp(z + \omega_1) = \wp(z)$. Similarly, $\wp(z + \omega_2) = \wp(z)$. □

It follows at once from the definition of compactness that any meromorphic function $f(z)$ has only finitely many zeros and poles in any bounded domain. Therefore it is possible to choose $\alpha \in \mathbb{C}$, such that the boundary ∂T_α of the translate

$$T_\alpha = \alpha + T = \{\alpha + z \mid z \in T\}$$

of T by α does not contain any zero or pole of $f(z)$.

THEOREM A.8. *Suppose $f(z) \in \mathbb{E}(L)$ and $f(z)$ has no zeros or poles on ∂T_α. If $z_1, \ldots, z_m$ are all the poles of $f(z)$ in T_α, then*

$$\sum_{j=1}^{m} \operatorname{Res}_{z_j} f = 0.$$

PROOF. By residue theorem, if we integrate $f(z)$ along ∂T_α counterclockwise, we obtain

$$\sum_{j=1}^{m} \operatorname{Res}_{z_j} f = \frac{1}{2\pi i} \int_{\partial T_\alpha} f(z)\, dz.$$

It is enough to show that the integral

$$\int_{\partial T_\alpha} f(z)\, dz = 0.$$

But this is obvious, because the values of $f(z)$ at the corresponding points on the opposite sides are equal, whereas the integration is along opposite directions.

THEOREM A.9. *Suppose a function $f(z)$ in $\mathbb{E}(L)$ has no zeros or poles on ∂T_α. Let m_j (or, respectively, n_j) be the order of various zeros (respectively, poles) of $f(z)$ in T. Then*

$$\sum m_j = \sum n_j.$$

PROOF. Let

$$f(z) = c_m(z-a)^m + c_{m+1}(z-a)^{m+1} + \cdots \qquad (c_m \neq 0)$$

be the Laurent expansion of $f(z)$ at $z = a$. Then

$$f'(z) = mc_m(z-a)^{m-1} + (m+1)c_{m+1}(z-a)^m + \cdots.$$

Therefore

1. If $m = 0$, $f'(z)/f(z)$ has no pole at $z = a$;
2. If $m \neq 0$, $f'(z)/f(z)$ has a simple pole at $z = a$.

Moreover, $\text{Res}_a f = m$. Hence by Theorem A.8,

$$\sum m_j - \sum n_j = 0. \qquad \square$$

If ∂T_α has no zeros or poles of $\wp(z)$ then $\wp(z)$ has exactly one double pole in the interior of T_α. The same is true of $\wp(z) - u$ for any complex number u. By Theorem A.9, $\wp(z) - u$ has exactly two zeros (or one double zero) in T_α. Thus we have proved the following theorem.

THEOREM A.10. *The Weierstrass function $\wp(z, L)$ assumes each value $u \in \mathbb{C}$ twice on the torus $T = \mathbb{C}/L$.*

A double zero $z = a$ of $\wp(z) - u$ is a zero of

$$\wp'(z) = -2 \sum_{\omega \in L - \{0\}} \frac{1}{(z - \omega)^3}.$$

Since $\wp(z)$ has exactly one double pole in T at $z = 0$ with principal part $1/z^2$, $\wp'(z)$ has exactly one triple pole there. So $\wp'(z)$ must have three zeros in T. In fact, the three zeros of $\wp'(z)$ are $\omega_1/2$, $\omega_2/2$, $(\omega_1 + \omega_2)/2$. To see this, note that $\wp'(z)$ is an odd function. Therefore

(1) For $j = 1, 2$

$$\begin{aligned} \wp'\left(\frac{\omega_j}{2}\right) &= \wp'\left(\frac{\omega_j}{2} - \omega_j\right) \\ &= \wp'\left(-\frac{\omega_j}{2}\right) \\ &= -\wp'\left(\frac{\omega_j}{2}\right) \end{aligned}$$

which gives $\wp'(\omega_j/2) = 0$.

(2)

$$\begin{aligned} \wp'\left(\frac{\omega_1 + \omega_2}{2}\right) &= \wp'\left[\frac{\omega_1 + \omega_2}{2} - (\omega_1 + \omega_2)\right] \\ &= \wp'\left(-\frac{\omega_1 + \omega_2}{2}\right) \\ &= -\wp'\left(\frac{\omega_1 + \omega_2}{2}\right), \end{aligned}$$

i.e.,

$$\wp'\left(\frac{\omega_1 + \omega_2}{2}\right) = 0.$$

We put

$$e_1 = \wp\left(\frac{\omega_1}{2}\right), \qquad e_2 = \wp\left(\frac{\omega_2}{2}\right)$$
$$e_3 = \wp\left(\frac{\omega_1 + \omega_2}{2}\right). \tag{A.6}$$

THEOREM A.11. *The complex numbers* e_1, e_2, e_3 *defined above are*

1. *The only values of* u *for which* $\wp(z) - u$ *has a double zero;*
2. *The only zeros of* $\wp'(z)$;
3. *All distinct.*

PROOF. It only remains to prove (3). Suppose $e_1 = e_2$. Then $\omega_1/2$, $\omega_2/2$ are both double zeros of $\wp(z) - e_1$ and consequently $\wp(z) - e_1$ has at least four zeros. This contradicts Theorem A.10. So $e_1 \neq e_2$. Similarly $e_2 \neq e_3$ and $e_3 \neq e_1$. □

A.3. The Weierstrass Equation

The most important property of the Weierstrass $\wp$-function is that it makes it possible to identify an elliptic curve E, or rather $E(\mathbb{C})$, with a torus $T = \mathbb{C}/L$.

THEOREM A.12. *For each lattice* L, *there are constants* $g_2 = g_2(L)$, $g_3 = g_3(L)$ *in* $\mathbb{C}$ *such that*

1. $\Delta = g_2^3 - 27g_3^2 \neq 0$;
2. *For all nonzero* z *in* $T = \mathbb{C}/L$, *we have the Weierstrass equation*

$$[\wp'(z, L)]^2 = 4\wp(z, L)^3 - g_2\wp(z, L) - g_3.$$

PROOF. We know that any elliptic function on T without poles is a constant function. So all we need is a cubic polynomial $f(x) = ax^3 + bx^2 + cx + d \in \mathbb{C}[x]$ of nonzero discriminant, such that the elliptic function

$$h(z) = \wp'(z)^2 - f(\wp(z))$$

has no pole in T and is zero for one z in T.

The only pole of $\wp(z)$ in T is a double pole at $z = 0$ with its principal part $1/z^2$, and thus the only possible pole of $h(z)$ in T is at $z = 0$. So we only need to compute the principal part of $h(z)$ at $z = 0$.

First we rewrite $\wp(z)$ differently. If we stay in a small enough neighborhood U of zero, then $|z/\omega| < 1$ for all nonzero ω in L. If $u \in \mathbb{C}$, $|u| < 1$, by differentiating

$$\frac{1}{1-u} = 1 + u + u^2 + u^3 + \cdots,$$

we obtain

$$\frac{1}{(1-u)^2} = 1 + 2u + 3u^2 + 4u^3 + \cdots.$$

So in U,

$$\begin{aligned}\frac{1}{(z-\omega)^2} &= \frac{1}{\omega^2} \cdot \frac{1}{(1 - z/\omega)^2} \\ &= \frac{1}{\omega^2}\left(1 + 2\frac{z}{\omega} + 3\frac{z^2}{\omega^2} + 4\frac{z^3}{\omega^3} + \cdots\right).\end{aligned}$$

Therefore

$$\begin{aligned}\frac{1}{(z-\omega)^2} - \frac{1}{\omega^2} &= 2\frac{z}{\omega^3} + 3\frac{z^2}{\omega^4} + 4\frac{z^3}{\omega^5} + \cdots \\ &= \sum_{j=3}^{\infty} (j-1)\frac{z^{j-2}}{\omega^j}.\end{aligned}$$

Since the series

$$\sum_{\omega \in L - \{0\}} \left[\frac{1}{(z-\omega)^2} - \frac{1}{\omega^2}\right] = \sum_{\omega \in L-\{0\}} \sum_{j=3}^{\infty} (j-1)\frac{z^{j-2}}{\omega^j}$$

converges absolutely, we may rearrange its terms as we like. In particular, by interchanging the order of summation, we obtain

$$\wp(z) = \frac{1}{z^2} + \sum_{j=3}^{\infty} (j-1) z^{j-2} \sum_{\omega \in L-\{0\}} \frac{1}{\omega^j}.$$

Recall that the series

$$\sum_{\omega \in L-\{0\}} \frac{1}{\omega^j} \qquad (j > 2)$$

is the Eisenstein series $G_j = G_j(L)$ relative to L and that $G_j = 0$ for j odd. Hence,

$$\begin{aligned}\wp(z) &= \frac{1}{z^2} + \sum_{k=2}^{\infty} (2k-1) G_{2k} z^{2k-2} \\ &= \frac{1}{z^2} + 3G_4 z^2 + 5G_6 z^4 + 7G_8 z^6 + \cdots. \qquad \text{(A.7)}\end{aligned}$$

From this we obtain

$$\wp(z)^2 = \frac{1}{z^4} + 6G_4 + 10G_6 z^2 + \cdots$$

and

$$\wp(z)^3 = \frac{1}{z^6} + 9G_4 \frac{1}{z^2} + 15G_6 + (21G_8 + 27G_4^2) z^2 + \cdots.$$

Since the expansion (A.7) for $\wp(z)$ is absolutely convergent, we may differentiate it termwise to obtain

$$\wp'(z) = \frac{-2}{z^3} + 6G_4 z + 20G_6 z^3 + 42G_8 z^5 + \cdots,$$

so that

$$\wp'(z)^2 = \frac{4}{z^6} - 24G_4 \frac{1}{z^2} - 80G_6 + (36G_4^2 - 168G_8) z^2 + \cdots.$$

Now it can be easily checked that

$$\begin{aligned}h(z) &= \wp'(z)^2 - a\wp(z)^3 - b\wp(z)^2 - c\wp(z) - d \\ &= \frac{A}{z^6} + \frac{B}{z^4} + \frac{C}{z^2} + D + a_2 z^2 + a_4 z^4 + \cdots,\end{aligned}$$

where

$$\begin{aligned}A &= a - 4, \\ B &= -b, \\ C &= -(24 + 9a)G_4 - c, \\ D &= -(80 + 15a)G_6 - 6bG_4 - d.\end{aligned}$$

Therefore the principal part of $h(z)$ is zero and $h(0) = 0$, provided $A = B = C = D = 0$, i.e., if

$$a = 4, \qquad b = 0, \qquad c = -60G_4, \qquad d = -140G_6.$$

Hence, if we define g_2 and g_3 in terms of the Eisenstein series $G_4(L)$ and $G_6(L)$ of weight four and six, respectively, by

$$g_2 = g_2(L) = 60G_4(L),$$

and (A.8)

$$g_3 = g_3(L) = 140G_6(L),$$

then the function

$$h(z) = \wp'(z)^2 - 4\wp(z)^3 + g_2\wp(z) + g_3$$

is identically zero.

It is clear now that for any lattice L, the three zeros of $4\wp(z)^3 - g_2\wp(z) - g_3$ are the three (distinct) zeros e_1, e_2, e_3 of $\wp'(z)$ defined by (A.6). So the discriminant Δ of $4x^3 - g_2x - g_3$ satisfies the condition

$$\Delta = g_2^3 - 27g_3^2 \neq 0. \qquad \square$$

Thus g_2, g_3 given by (A.8) define an elliptic curve

$$E_L\colon y^2 = 4x^3 - g_2x - g_3.$$

Note that the above equation can also be written as

$$y^2 = 4(x - e_1)(x - e_2)(x - e_3).$$

We now study the map

$$\Phi = \Phi_L\colon \mathbb{C}/L \to E_L(\mathbb{C}) \subseteq \mathbb{P}^2(\mathbb{C})$$

given by

$$\Phi(z) = \begin{cases} (\wp(z), \wp'(z), 1) & \text{if } z \neq 0, \\ O, \quad \text{the point at infinity} & \text{if } z = 0. \end{cases}$$

This map is analytic, i.e., in a neighborhood U of any point $z_0 \in \mathbb{C}/L$, the coordinates of the point $\Phi(z)$ on $E(\mathbb{C})$ are analytic functions on U. This is obvious at $z_0 \neq 0$. At $z_0 = 0$ also $\Phi(z)$ is analytic, because in $\mathbb{P}^2(\mathbb{C})$, we may write

$$\Phi(z) = (\wp(z)/\wp'(z), 1, 1/\wp'(z)).$$

THEOREM A.13. *The analytic map Φ_L is a bijection between $\mathbb{C}/L$ and $E_L(\mathbb{C})$.*

PROOF. We have seen (Theorem A.10) that the function

$$\wp\colon \mathbb{C}/L \to \mathbb{C} \cup \{\infty\}$$

is a two-to-one and onto function. In fact, unless $z = -z$, i.e., unless $2z = 0$, we have $\wp(z) = \wp(-z)$ and $\wp'(-z) = -\wp'(z) \neq 0$. On the other hand, corresponding to each $x \in \mathbb{C}$, there are two points (x, y), $(x, -y)$ on $E_L(\mathbb{C})$, unless $y = 0$. Thus except for $\{z \in \mathbb{C}/L \mid 2z = 0\}$, i.e., except for $z = 0, \omega_1/2, \omega_2/2, (\omega_1 + \omega_2)/2$, the point z (or, respectively $-z$) is mapped to $(\wp(z), \wp'(z))$ [respectively, $(\wp(z), -\wp'(z))$]. So $\Phi(z) \neq \Phi(-z)$. Finally $\wp(0)$, $=\infty$, so that $\Phi(0) = O$ and corresponding to three distinct values e_1, e_2, e_3 of $x = \wp(z)$ at $z = \omega_1/2, \omega_2/2, (\omega_1 + \omega_2)/2$, we have $y = \wp'(z) = 0$. This proves that Φ is one-to-one.

REMARK A.14. Since Φ is bijective, its inverse Φ^{-1} exists. This map Φ^{-1} is also analytic. To see this fix a point z_0 on T. Solve the differential equation

$$\left(\frac{d\wp}{dz}\right)^2 = 4\wp^3 - g_2\wp - g_3$$

to obtain

$$z = \int_{\wp(z_0)}^{\wp(z)} \frac{d\wp}{(4\wp^3 - g_2\wp - g_3)^{1/2}} + z_0.$$

The path is taken along a curve avoiding the singularities of the integrand and the branch of the square root function is chosen properly.

A.4. Addition Theorems

We have seen that $\mathbb{C}/L$ and $E(\mathbb{C})$ can be identified as "analytic manifolds" via the map Φ. We now show that $\mathbb{C}/L$ and $E(\mathbb{C})$ are isomorphic groups.

THEOREM A.15. *The analytic map*

$$\Phi_L: \mathbb{C}/L \to E_L(\mathbb{C})$$

is a bijective group homomorphism.

PROOF. Since Φ_L is bijective (Theorem A.13), it is enough to prove the following two identities, called the *addition theorems*:

1. If $z_1, z_2 \in T = \mathbb{C}/L$ and $z_1 \neq \pm z_2$, then

$$\wp(z_1 + z_2) = \frac{1}{4}\left[\frac{\wp'(z_1) - \wp'(z_2)}{\wp(z_1) - \wp(z_2)}\right]^2 - \wp(z_1) - \wp(z_2); \qquad \text{(A.9)}$$

and

2. If $z \in T$ such that $2z \neq 0$, then

$$\wp(2z) = \frac{1}{4}\left[\frac{\wp''(z)}{\wp'(z)}\right]^2 - 2\wp(z).$$

Since part (2) is the limit of part (1) as $z_2 \to z_1 = z$, it suffices to prove (1) only.

We keep z_1 fixed and suppose that $z_1 \neq 0, \omega_1/2, \omega_2/2, (\omega_1 + \omega_2)/2$. Let $z = z_2$ vary and define an elliptic function $h(z)$ on T by

$$h(z) = \frac{1}{4}\left[\frac{\wp'(z) - \wp'(z_1)}{\wp(z) - \wp(z_1)}\right]^2 - \wp(z) - \wp(z_1) - \wp(z + z_1).$$

We show that $h(z)$ is identically zero.

The only possible poles of the elliptic function $h(z)$ are at $z = 0$ and at $z = -z_1$. At $z = 0$, we have the following Laurent expansions:

$$\wp(z) = \frac{1}{z^2} + a_2 z^2 + a_4 z^4 + \cdots,$$

$$\wp'(z) = \frac{-2}{z^3} + 2a_2 z + 4a_4 z^3 + \cdots.$$

Therefore

$$\begin{aligned}\frac{1}{4}\left[\frac{\wp'(z) - \wp'(z_1)}{\wp(z) - \wp(z_1)}\right]^2 - \wp(z) &= \frac{1}{z^2}\left(\frac{1 + \frac{1}{2}\wp'(z_1)z^3 - a_2 z^4 + \cdots}{1 - \wp(z_1)z^2 + a_2 z^4 + \cdots}\right)^2 \\ &\quad - \left(\frac{1}{z^2} + a_2 z^2 + a_4 z^4 + \cdots\right) \\ &= \frac{1}{z^2}[1 + \wp'(z_1)z^3 + \cdots][1 + 2\wp(z_1)z^2 + \cdots] \\ &\quad - \left(\frac{1}{z^2} + a_2 z^2 + a_4 z^4 + \cdots\right) \\ &= 2\wp(z_1) + b_1 z + \cdots.\end{aligned}$$

This shows that $h(z)$ is holomorphic at $z = 0$ with $h(0) = 0$.

At $z = -z_1$, consider the following Taylor expansions:

$$\wp'(z) = -\wp'(z_1) + \wp''(z_1)(z + z_1) - \tfrac{1}{2}\wp'''(z_1)(z + z_1)^2 + \cdots$$

and

$$\wp(z) = \wp(z_1) - \wp'(z_1)(z + z_1) + \tfrac{1}{2}\wp''(z_1)(z + z_1)^2 + \cdots.$$

By assumption, $\wp'(z_1) \neq 0$, so using the above expansions and putting

$$\alpha = \frac{\wp''(z_1)}{\wp'(z_1)}$$

we obtain

$$\frac{1}{4}\left[\frac{\wp'(z)-\wp'(z_1)}{\wp(z)-\wp(z_1)}\right]^2 = \frac{1}{(z+z_1)^2}[1-\alpha(z+z_1)+\cdots][1+\alpha(z+z_1)+\cdots]$$
$$= \frac{1}{(z+z_1)^2} + \text{terms of degree} \geq 0.$$

This and the expansion

$$\wp(z+z_1) = \frac{1}{(z+z_1)^2} + \text{terms of degree} \geq 0$$

show that $h(z)$ is holomorphic at $z = -z_1$ also. This proves that $h(z)$ is identically zero.

Finally the role of z_1 and z_2 can be interchanged unless z_1, z_2 are both in the set $S = \{0, \omega_1/2, \omega_2/2, (\omega_1+\omega_2)/2\}$. Since S is *discrete* (each point of S is isolated), by continuity, (A.9) must hold if $z_1 \neq \pm z_2$.

A.5. Isomorphic Classes of Elliptic Curves

If

$$f: G_1 \to G_2$$

is a homomorphism of abelian groups G_1, G_2 with subgroups H_1, H_2, respectively, such that $f(H_1) = \{f(h) \mid h \in H_1\} \subseteq H_2$, then f *induces* a homomorphism

$$\bar{f}: G/H_1 \to G/H_2$$

of the quotient groups. It is given by

$$\bar{f}(h + H_1) = f(h) + H_2.$$

Remark A.16. It is obvious that

1. If f is onto, then so is $\bar{f}$;

and

2. $\bar{f}$ is one-to-one if and only if

$$f^{-1}(H_2) = \{g \in G_1 \mid f(g) \in H_2\} = H_1.$$

THEOREM A.17. *Suppose L_1, L_2 are two lattices in $\mathbb{C}$ and*

$$f\colon \mathbb{C}/L_1 \to \mathbb{C}/L_2 \tag{A.10}$$

is an analytic group homomorphism. Then f is induced by the multiplication map

$$m_\lambda\colon \mathbb{C} \to \mathbb{C}$$

given by $m_\lambda(z) = \lambda z$, for some $\lambda \in \mathbb{C}$. In particular, $\lambda L_1 = \{\lambda\omega \mid \omega \in L_1\} \subseteq L_2$.

PROOF. At $z = 0$, we have the Taylor expansion

$$f(z) = a_0 + a_1 z + a_2 z^2 + a_3 z^3 + \cdots.$$

Since f is a group homomorphism, $f(0) = 0$ which shows that $a_0 = 0$ and for small z,

$$f(2z) = 2f(z) = \sum_{n=1}^{\infty} 2a_n z^n.$$

But evaluating the above Taylor expansion of $f(z)$ at $2z$,

$$f(2z) = \sum_{n=1}^{\infty} 2^n a_n z^n.$$

Equating the coefficients, it follows at once that $a_n = 0$ for all $n > 1$. Thus near zero, $f(z)$ is multiplication by $\lambda = a_1$. For arbitrary z, choose n large enough so that z/n is near zero. Because f is a group homomorphism,

$$f(z) = f\left(n \cdot \frac{z}{n}\right) = nf\left(\frac{z}{n}\right) = n \cdot \lambda \cdot \frac{z}{n} = \lambda z. \qquad \square$$

By Remark A.16, the homomorphism (A.10) is an isomorphism if and only if $\lambda L_1 = L_2$.

DEFINITION A.18. Let k be a subfield of $\mathbb{C}$. Two lattices L, L' are *linearly equivalent over k*, if $L' = \lambda L$ for some $\lambda \in k$, i.e., if $m_\lambda\colon \mathbb{C}/L \to \mathbb{C}/L'$ is an isomorphism with $\lambda \in k$.

DEFINITION A.19. Two elliptic curves E, E' defined by

$$y^2 = 4x^3 - g_2 x - g_3$$

and

$$y^2 = 4x^3 - g'_2 x - g'_3$$

respectively, are said to be *isomorphic over k*, if for some $\mu \in k$, we can write

$$g'_2 = \mu^4 g_2, \qquad g'_3 = \mu^6 g_3. \tag{A.11}$$

Note that both λ and μ are necessarily nonzero.

THEOREM A.20. *If L is linearly equivalent to L' over k by $\lambda \in k$, i.e., $L' = \lambda L$, then E_L is isomorphic to $E_{L'}$ over k with $\mu = 1/\lambda$ in* (A.11). *Moreover, the isomorphism m_λ: $\mathbb{C}/L \to \mathbb{C}/L'$ induces an isomorphism $\phi_\lambda = \Phi_{L'} \circ m_\lambda \circ \Phi_L^{-1}$: $E_L(\mathbb{C}) \to E_{L'}(\mathbb{C})$ given by*

$$(x', y') = \phi_\lambda(x, y) = (\mu^2 x, \mu^3 y).$$

PROOF. Since $L' = \lambda L$, it is clear that

$$\begin{aligned} g_2' &= g_2(L') = 60G_4(L') \\ &= 60 \sum_{\omega' \in L' - \{0\}} \frac{1}{\omega'^4} = 60 \sum_{\omega \in L - \{0\}} \frac{1}{(\lambda\omega)^4} = \frac{60}{\lambda^4} G_4(L) = \frac{g_2}{\lambda^4}. \end{aligned}$$

Similarly $g_3' = g_3/\lambda^6$. Moreover, by definition,

$$x' = \wp(\lambda z, \lambda L) = \frac{1}{\lambda^2}\wp(z, L) = \frac{x}{\lambda^2}$$

and

$$y' = \wp'(\lambda z, \lambda L) = \frac{1}{\lambda^3}\wp'(z, L) = \frac{y}{\lambda^3}.$$

EXAMPLE A.21. Let $i = \sqrt{-1}$ and consider the lattice $L = \mathbb{Z}[i] = \mathbb{Z} \oplus \mathbb{Z}i = \{m + ni \mid m, n \in \mathbb{Z}\}$. In fact L is a ring, called the *ring of gaussian integers*. Clearly $iL = L$. So

$$g_3(L) = g_3(iL) = i^{-6} g_3(L) = -g_3(L).$$

This shows that $g_3(L) = 0$ and E_L is given by the equation

$$y^2 = 4x^3 - g_2 x. \tag{A.12}$$

Now for any $g_2' \in \mathbb{C}^\times$, choosing $\lambda = (g_2/g_2')^{1/4}$, we obtain

$$g_2(\lambda L) = \frac{1}{\lambda^4} g_2(L) = g_2'.$$

Thus any elliptic curve of the form (A.12) is E_L for some lattice L.

EXAMPLE A.22. Suppose ω is a primitive cube root of unity, e.g.,

$$\omega = \frac{-1 + \sqrt{-3}}{2}.$$

Let $L = \mathbb{Z}[\omega] = \mathbb{Z} \oplus \mathbb{Z}\omega$. Again $\omega L = L$ and as before,

$$\begin{aligned} g_2(L) &= g_2(\omega L) \\ &= \omega^2 g_2(L). \end{aligned}$$

Since $\omega^2 - 1 \neq 0$, $g_2(L) = 0$ and E_L is

$$y^2 = 4x^3 - g_3. \tag{A.13}$$

And as before we see that any elliptic curve defined by an equation of the type (A.13) is E_L for some L.

REMARK A.23. We have associated to a class of linearly equivalent lattices a unique class of isomorphic elliptic curves. Conversely, it is also true that each isomorphic class of elliptic curves is associated to a class of linearly equivalent lattices. We have proved the converse only in the special case when g_2 or $g_3 = 0$. For the proof in the general case when $g_2, g_3 \in \mathbb{C}^\times$ with $\Delta = g_2^3 - 27g_3^2 \neq 0$, the theory of modular functions comes in. See Theorem 6, Chapter VI of Ref. 2.

A.6. Endomorphisms of an Elliptic Curve

Now we fix an elliptic curve E, which we may take to be $\mathbb{C}/L$ for a lattice L. An *endomorphism* of E is an analytic homomorphism of $\mathbb{C}/L$ into itself. It is the multiplication

$$m_\lambda : \mathbb{C}/L \to \mathbb{C}/L$$

by a complex number $\lambda \in \mathbb{C}$ such that $\lambda L \subseteq L$. We put

$$\mathrm{End}(E) = \{\lambda \in \mathbb{C} \mid \lambda L \subseteq L\}.$$

Clearly $\mathrm{End}(E)$ is a ring containing $\mathbb{Z}$. Let if possible $\mathrm{End}(E) \supsetneqq \mathbb{Z}$. Suppose $\lambda \in \mathbb{C} - \mathbb{Z}$. Since $\lambda L \subseteq L = \mathbb{Z}\omega_1 \oplus \mathbb{Z}\omega_2$,

$$\begin{aligned} \lambda\omega_1 &= a\omega_1 + b\omega_2 \\ \lambda\omega_2 &= c\omega_1 + d\omega_2 \end{aligned} \tag{A.14}$$

for integers a, b, c, d. This shows that λ is a root of the monic polynomial equation

$$\begin{vmatrix} x - a & -b \\ -c & x - d \end{vmatrix} = 0$$

of degree two.

Recall that $\tau = \omega_1/\omega_2$ is not real, so dividing the second equation in (A.14) by ω_2, we obtain

$$\lambda = c\tau + d.$$

This shows that λ is a nonreal algebraic integer of degree two over $\mathbb{Q}$ and $\mathbb{Q}(\lambda) = \mathbb{Q}(\tau)$. Hence *any* $\lambda \in \mathbb{C}$ *such that* $\lambda L \subseteq L$ *must be in the ring* $\mathcal{O}_k$ *of integers of the fixed imaginary quadratic field* $k = \mathbb{Q}(\tau)$, i.e., $\mathrm{End}(E) \subseteq \mathcal{O}_k$. This is the justification for the following definition:

DEFINITION A.24. Suppose E is an elliptic curve. We say that E has *complex multiplication* if $\mathbb{Z} \subsetneqq \mathrm{End}(E)$.

Equations (A.12) and (A.13) provide some examples of elliptic curves with complex multiplication. In fact, for these elliptic curves, $\mathrm{End}(E) = \mathbb{Z}[i]$ and $\mathbb{Z}[\omega]$, respectively.

A bijective endomorphism of E is called an *automorphism* of E. The (multiplicative) groups $\mathrm{Aut}(E)$ of automorphism of E is a subgroup of the group $\mathcal{O}_k^\times$ of units of k. Because k is an imaginary quadratic field, it is easy to check (use Dirichlet's theorem) that

$$\mathcal{O}_k^\times = \begin{cases} \langle i \rangle & \text{if } k = \mathbb{Q}(\sqrt{-1}), \\ \langle -\omega \rangle & \text{if } k = \mathbb{Q}(\sqrt{-3}), \\ \langle -1 \rangle & \text{otherwise,} \end{cases}$$

where $\omega = (-1 + \sqrt{-3})/2$ and $\langle x \rangle$ denote the subgroup generated by an element x of a group G.

Alternately, an automorphism ϕ of E, where E now is given by say

$$y^2 = 4x^3 - g_2 x - g_3,$$

is of the type $\phi(x, y) = (\mu^2 x, \mu^3 y)$ with $\mu^4 g_2 = g_2$ and $\mu^6 g_3 = g_3$. Thus the only automorphisms of E are as follows:

1. If $g_2 g_3 \neq 0$, then $\phi(x, y) = (x, \pm y)$;
2. If $g_2 \neq 0$, $g_3 = 0$, then $\phi(x, y) = (x, \pm y)$ or $(-x, \pm iy)$;
3. If $g_2 = 0$, $g_3 \neq 0$, then $\phi(x, y) = (\omega x, \pm y)$, where ω is a cube root of unity.

A.7. Points of Finite Order

An important consequence of identifying an elliptic curve E with the torus $\mathbb{C}/L$ for a lattice L (Theorem A.15) is that it becomes trivial to determine the group of torsion points on E. In fact, for each integer $N \geq 1$,

$$\begin{aligned} E[N] &= \{P \in E(\mathbb{C}) \mid NP = O\} \\ &\cong \{z \in T = \mathbb{C}/L \mid Nz = 0\} \\ &= \langle \omega_1/N \rangle \oplus \langle \omega_2/N \rangle \\ &\cong \mathbb{Z}/N\mathbb{Z} \times \mathbb{Z}/N\mathbb{Z}. \end{aligned}$$

In particular, if E is defined over a number field k and K is a subfield of $\mathbb{C}$ containing k, then $E(K)[N]$ is a subgroup of $\mathbb{Z}/N\mathbb{Z} \times \mathbb{Z}/N\mathbb{Z}$. Once we have identified $E[N]$ with $\mathbb{Z}/N\mathbb{Z} \times \mathbb{Z}/N\mathbb{Z}$, $E[N]$ becomes a module over the ring $\mathbb{Z}/N\mathbb{Z}$.

Now suppose that E is an elliptic curve defined over a number field k. We shall show that *if $P = (x, y)$ is a point of finite order, then x, y are necessarily algebraic over k.* Let $K = k(E[N])$ denote the field obtained by adjoining to k the coordinates of N^2 points of $E[N]$. We call K the *field of N-division points of E over k.* Let L be a subfield of $\mathbb{C}$ containing K and $\sigma: L \to \mathbb{C}$ an isomorphism of fields over k. For each $P = (x, y) \in E(L)$, it is clear that

$$\sigma(P) = (\sigma(x), \sigma(y))$$

is in $E(\sigma(L))$. Moreover, for P, Q in E, the coordinates of $P + Q$ are the rational functions (with coefficients in k) of the coordinates of P and Q. Hence

$$\sigma(P + Q) = \sigma(P) + \sigma(Q).$$

In particular, if $P \in E[N]$, then

$$N\sigma(P) = \sigma(NP) = \sigma(O) = O$$

shows that $\sigma(P)$ is also in $E[N]$. In other words, σ permutes the finite set of numbers consisting of the coordinates of points in $E[N]$.

THEOREM A.25. *The field $K = K_N = k(E[N])$ of N-division points of E over k is a (finite) galois extension of k.*

PROOF. It is obvious that K/k is normal, because for each isomorphism $\sigma: K \to \mathbb{C}$ that leaves the elements of the ground field k fixed, $\sigma(K) \subseteq K$. To show that K/k is finite it suffices to show that if $P = (x, y) \in E[N]$, then x, y are algebraic numbers. This follows at once from the following simple observation.

If $\alpha_1, \ldots, \alpha_n \in \mathbb{C}$ such that for any subfield L of $\mathbb{C}$ containing the field $k(\alpha_1, \ldots, \alpha_n)$, every isomorphism $\sigma: L \to \mathbb{C}$ over k permutes $\alpha_1, \ldots, \alpha_n$, then $\alpha_1, \ldots, \alpha_n$ are algebraic numbers. □

Let P_1, P_2 correspond to $(1, 0)$ and $(0, 1)$, respectively, under the identification of $E[N]$ with $\mathbb{Z}/N\mathbb{Z} \times \mathbb{Z}/N\mathbb{Z}$. Then P_1, P_2 is a basis of the $\mathbb{Z}/N\mathbb{Z}$-module $E[N]$. If $\sigma \in \mathrm{Gal}(K_N/k)$, then

$$\left.\begin{aligned} \sigma(P_1) &= aP_1 + cP_2 \\ \sigma(P_2) &= bP_1 + dP_2 \end{aligned}\right\} \qquad (a, b, c, d \in \mathbb{Z}/N\mathbb{Z}),$$

defines a group homomorphism

$$\rho_N\colon G_N = \mathrm{Gal}(K_N/k) \to \mathrm{GL}_2(\mathbb{Z}/N\mathbb{Z}), \tag{A.15}$$

called a *representation* of G_N into the general linear group $\mathrm{GL}_2(\mathbb{Z}/N\mathbb{Z})$ of 2×2 invertible matrices over the ring $\mathbb{Z}/N\mathbb{Z}$.

If

$$\rho_N(\sigma) = I = \begin{pmatrix} 1 & 0 \\ 0 & 1 \end{pmatrix},$$

then $\sigma(P_j) = P_j (j = 1, 2)$, which implies that $\sigma(P) = P$ for all $P \in E[N]$. Hence σ is identity on K_N. This shows that ρ_N is injective and G_N can be considered as a subgroup of $\mathrm{GL}_2(\mathbb{Z}/N\mathbb{Z})$.

Recall the remark (Section 4.8.2) that the homomorphism

$$\Phi_m\colon \mathrm{Gal}(\mathbb{Q}(\zeta_m)/\mathbb{Q}) \to (\mathbb{Z}/m\mathbb{Z})^\times$$

is onto. A basic problem in number theory is this: How close is the homomorphism (A.15) to being surjective? In other words, how small is the index $[\mathrm{GL}_2(\mathbb{Z}/N\mathbb{Z}) : \rho_N(G_N)]$? We shall mention only one result (which provides further examples of non-abelian extensions). For details, see Ref. 3.

THEOREM A.26* (*Serre* [3]). *If E has no complex multiplication, there is a constant $c > 0$, depending on E and k only such that for all $N \in \mathbb{N}$,*

$$[\mathrm{GL}_2(\mathbb{Z}/N\mathbb{Z}) \colon \rho_N(G_N)] \le c.$$

References

1. L. V. Ahlfors, *Complex Analysis*, 3rd Edition, McGraw-Hill, New York (1979).
2. K. C. Chandrasekharan, *Elliptic Functions*, Grundlehren der Mathematischen Wissenschaften 281, Springer Verlag, Berlin, (1985).
3. J.-P. Serre, Propriétés galoisiennes des points d'order fini des courbes elliptiques, *Invent. Math.* **15**, 229–331 (1972).

Some Great Number Theorists

Pythagoras (572–492 B.C.)
Euclid (c.303–c.275 B.C.)
Diophantus (c.246–c.330 A.D.)
Pierre de Fermat (1601–1665)
Leonhard Euler (1707–1783)
Joseph Louis Lagrange (1736–1813)
Adrian Marie Legendre (1752–1833)
Carl Friedrich Gauss (1777–1855)
Peter Gustav Lejeune Dirichlet (1805–1859)
Evariste Galois (1811–1832)
Karl Wilhelm Theodor Weierstrass (1815–1897)
Charles Hermite (1822–1901)
Ferdinand Gottfried Max Eisenstein (1823–1852)
Leopold Kronecker (1823–1891)
Georg Friedrich Bernhard Riemann (1826–1866)
Heinrich Weber (1842–1913)
Jules Henri Poincaré (1854–1912)
Adolf Hurwitz (1859–1919)
David Hilbert (1862–1943)
Louis Joel Mordell (1888–1972)
Carl Ludwig Siegel (1896–1981)
Emil Artin (1898–1962)
Helmut Hasse (1898–1979)
André Weil (1906–)

Index